DELAY-DOPPLER COMMUNICATIONS

DELAY-DOPPLER COMMUNICATIONS

Principles and Applications

Yi Hong

Department of Electrical and Computer Systems Engineering
Monash University
Melbourne, VIC, Australia

Tharaj Thaj

Department of Electrical and Computer Systems Engineering
Monash University
Melbourne, VIC, Australia

Emanuele Viterbo

Department of Electrical and Computer Systems Engineering
Monash University
Melbourne, VIC, Australia

ELSEVIER

ACADEMIC PRESS

An imprint of Elsevier

Academic Press is an imprint of Elsevier
125 London Wall, London EC2Y 5AS, United Kingdom
525 B Street, Suite 1650, San Diego, CA 92101, United States
50 Hampshire Street, 5th Floor, Cambridge, MA 02139, United States
The Boulevard, Langford Lane, Kidlington, Oxford OX5 1GB, United Kingdom

Library of Congress Cataloging-in-Publication Data
A catalog record for this book is available from the Library of Congress

British Library Cataloguing-in-Publication Data
A catalogue record for this book is available from the British Library

ISBN: 978-0-323-85028-5

For information on all Academic Press publications
visit our website at https://www.elsevier.com/books-and-journals

Publisher: Mara Conner
Acquisitions Editor: Tim Pitts
Editorial Project Manager: Chiara Giglio
Production Project Manager: Kamesh Ramajogi
Designer: Greg Harris

Typeset by VTeX

We dedicate this work to our parents

Guomei, Guifa, Mariamma, Thajudeen, Mariella, and Davide

Contents

4. Delay-Doppler modulation

5. Zak transform analysis for delay-Doppler communications

Please visit the book's companion site for additional materials (MATLAB package code): https://www.elsevier.com/books-and-journals/book-companion/9780323850285

List of figures

Biography

Yi Hong

Dr. Yi Hong is an Associate Professor at the Department of Electrical and Computer Systems Engineering, Monash University, Australia. She obtained her PhD in Electrical Engineering and Telecommunications from the University of New South Wales (UNSW), Sydney, and received the NICTA-ACoRN Earlier Career Researcher Award at the 2007 Australian Communication Theory Workshop, Adelaide. Yi Hong served as a member of the Australian Research Council College of Experts in 2018–2020. She was an Associate Editor for *IEEE Wireless Communication Letters* and *Transactions on Emerging Telecommunications Technologies (ETT).* She was the Tutorial Chair of the 2021 IEEE International Symposium on Information Theory, held in Melbourne, and the General Co-Chair of the 2021 IEEE International Conference on Communications Workshop on Orthogonal Time Frequency Space Modulation (OTFS) for 6G and Future High-mobility Communications, held in Montreal. She was the General Co-Chair of the 2014 IEEE Information Theory Workshop, held in Hobart; the Technical Program Committee Chair of the 2011 *Australian Communications Theory Workshop*, held in Melbourne; and the 2009 Publicity Chair at the *IEEE Information Theory Workshop*, held in Sicily. Her research interests include communication theory, coding, and information theory with applications to telecommunication engineering.

Tharaj Thaj

Mr. Tharaj Thaj received his B.Tech. degree in Electronics and Communication Engineering from the National Institute of Technology, Calicut, India, in 2012 and his M.Tech. degree in Telecommunication Systems Engineering from the Indian Institute of Technology, Kharagpur, India, in 2015. He is currently pursuing a PhD with the Department of Electrical and Computer Systems Engineering, Monash University, Australia. From 2012 to 2013, he worked at Verizon Data Services India, as a Software Engineer, focusing on network layer routing algorithms and protocols. From 2015 to 2017, he worked as a Senior Engineer in the Communication, Navigation, and Surveillance (CNS) Department of Honeywell Technology Solutions Lab, Bengaluru. His current research interests include physical layer design and implementation of wireless communication systems for next-generation wireless networks.

Emanuele Viterbo

Dr. Emanuele Viterbo is a Professor in the Department of Electrical and Computer Systems Engineering at Monash University, Australia. He served as Head of Department and Associate Dean Graduate Research in the Faculty of Engineering at Monash University.

Prof. Viterbo obtained his degree and a PhD in Electrical Engineering, both from the Politecnico di Torino, Turin, Italy. From 1990 to 1992, he worked at the European Patent Office, The Hague, The Netherlands, as a patent examiner in the field of dynamic recording and error-control coding. Between 1995 and 1997, he held a post-doctoral position at Politecnico di Torino. In 1997–1998, he was a post-doctoral research fellow in the Information Sciences Research Center of AT&T Research, Florham Park, NJ, USA. He later joined the Dipartimento di Elettronica at Politecnico di Torino. From 2006 to August 2010, he was a Full Professor in DEIS at the University of Calabria, Italy. In September 2010, he joined the ECSE Department at Monash University as a Professor, where he is continuing his research.

Prof. Viterbo is a Fellow of the IEEE, an ISI Highly Cited Researcher, and a Member of the Board of Governors of the IEEE Information Theory Society (2011–2013 and 2014–2016). He served as an Associate Editor for *IEEE Transactions on Information Theory*, *European Transactions on Telecommunications*, and the *Journal of Communications and Networks*. His main research interests are in lattice codes for Gaussian and fading channels, algebraic coding theory, algebraic space-time coding, digital terrestrial television broadcasting, and digital magnetic recording.

Preface

Orthogonal frequency division multiplexing (OFDM) has been the waveform of choice for most wireless communications systems in the past 25 years. "What comes next?" – The book will address this question by illustrating the current limitations of OFDM when dealing with high-mobility environments and presenting the fundamentals of a new recently proposed waveform known as "orthogonal time-frequency space" (OTFS). The OTFS waveform is based on the idea that mobile wireless channels can be effectively modeled in the *delay-Doppler domain*. The information is encoded in such domain to combat the Doppler shifts in multipath propagation channels that are typically found in high-mobility environments.

This book has been developed from a number of tutorial presentations delivered by the authors at major conferences on wireless communications between 2018 and 2021. It expands and integrates several key research papers by the authors intending to provide a taxonomy for the different flavors of delay-Doppler communications.

We have chosen to present the fundamentals of delay-Doppler communications following two approaches: (i) The first one is oriented to readers who are familiar with multicarrier techniques and is based on the idea that the *delay-Doppler domain* and the *time-frequency domain* are related by a precoding operation, based on the *two-dimensional symplectic Fourier transform*. (ii) The second one is for readers interested in the underlying mathematical tools that provide a direct relation between the (two-dimensional) delay-Doppler domain and the (one-dimensional) time domain signals, effectively bypassing the time-frequency domain interpretation. This is based on the theory of the Zak transform and its specific properties.

We believe that both approaches should be followed and mastered by the reader interested in further developing delay-Doppler communications systems in high-mobility environments. As a first approach to the subject, we recommend Chapters 1–4 and 6 to cover the basic delay-Doppler modulation and demodulation. For readers interested in the Zak transform approach and the more specialized topics related to channel estimation, MIMO, and multiuser systems, Chapters 5, 7, and 8 will be of relevance. Finally, Chapter 9 touches upon some new research directions in high-mobility communications and outlines a more general two-dimensional scheme beyond the delay-Doppler.

We would like to express thanks and appreciation to our colleagues Dr. Raviteja Patchava (Qualcomm), Ezio Biglieri (Universitat Pompeu Fabra, Barcelona), Saif Mohammad Khan (IIT Delhi), and Ananthanarayanan Chokalingham (IISc. Bangalore), who have pioneered with us the interest in delay-Doppler communications. This book has taken shape by the many interactions and discussions with them. We thank Dr. Viduranga Wijekoon and Dr. Birenjith Sasidharan for their careful proofreading of our manuscript. A special thanks goes to the inventor of OTFS, Ronny Hadani (Cohere Technologies and University of Texas, Austin) for some very inspiring discussions with the authors. Since the first public disclosure of OTFS in 2017, he has helped us to appreciate the more abstract interpretation of OTFS.

Yi Hong, Tharaj Thaj, and Emanuele Viterbo
Monash University, Melbourne, VIC, Australia
September 2021

CHAPTER

1

Introduction

OUTLINE

The concept of cellular wireless networks emerged in the 1980s with the first generation analog radio communications systems (1G). These were an evolution of trunked radio systems that enabled personal wireless phones to become accessible to the general public. The radio spectrum was accessed using frequency division multiplexing (FDMA) and analog frequency modulation of the voice signals was used.

The second generation system (2G) *Global System for Mobile Communications* (GSM) was introduced in the early 1990s to embrace the benefits of digital communication technology, namely the possibility to compress and encrypt the voice signal and better control the quality of the communication in the presence of radio disturbances. This resulted in an increase in the number of users per frequency band, and some basic digital data communications became more accessible. A narrowband digital modulation for compressed digitized voice was adopted, and time division multiple access (TDMA) was used in conjunction with different frequency bands for channelization. The notion of time-frequency resources was beginning to emerge.

Prompted by the explosion of internet browsing, the third generation (3G) system was standardized at the turn of the century under the *International Telecommunications Union IMT-2000* standard comprising two variations: the Universal Mobile Telecommunications System (UMTS) and

Delay-Doppler Communications
https://doi.org/10.1016/B978-0-32-385028-5.00009-8

the code division multiple access 2000 (CDMA2000) system. Wideband signaling using the spread spectrum technique of CDMA was adopted to enable higher digital data rates and radically moved away from narrowband signaling in 2G. The CDMA scheme used Walsh–Hadamard spreading sequences as orthogonal basis functions, which effectively spread the information of different users in both time and frequency. The multiple access scheme for the uplink was realized by allowing these functions to overlap in both time and frequency and separating them at the receiver thanks to the different orthogonal codes (signatures). The channel impairments distort the signals and cause interference between users (multiple access interference) that needs to be compensated at the receiver. Severe multiple access interference hinders the quasi-orthogonality of the signature waveforms at the receiver and calls for more complex processing to recover the individual user information.

The above drawback led to the introduction of orthogonal frequency division multiple access (OFDMA) in 4G systems which was preserved in the 5G systems. Orthogonal frequency division multiplexing (OFDM) is a wideband signaling technique, where multiple information symbols overlapping in time are made orthogonal in the frequency domain by allocating them to suitably spaced subcarriers. It can be used for multiple access by allocating orthogonal time-frequency resources to the users. With OFDMA, orthogonality is maintained at the receiver even in the presence of some multipath channel impairments. Among the advantages of this solution, we have a relatively low processing cost for detection and channel estimation. Fig. 1.1 summarizes the evolution of the wireless systems.

As the mobility of terminals in the system increases, the immediate advantages of multicarrier technology tend to dissipate due to the rapidly time-varying nature of the channel affecting the orthogonality of the received signals and resulting in severe intersymbol interference. In order to continue using multicarrier techniques in such high-mobility environments, more complex equalization and more overhead for channel estimation are required.

In this book, we will illustrate how delay-Doppler communications can offer a solution to the limitations of multicarrier techniques over channels with high mobility, where Doppler shifts cannot be easily compensated for.

1.1 High-mobility wireless channels

In a *static wireless channel*, nothing in the environment is moving, and both transmitters and receivers are static. The electromagnetic signal, propagating from a transmit antenna, reaches the receiving antenna via

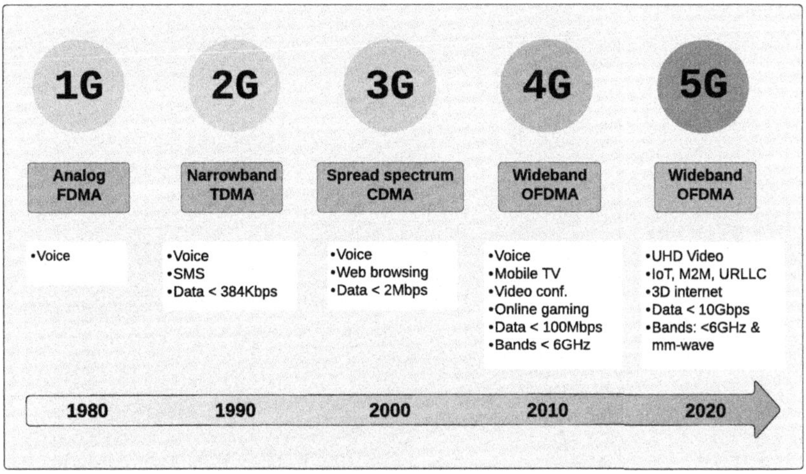

FIGURE 1.1 Evolution of the wireless systems.

multiple paths due to the presence of reflecting objects (scatterers). The receiver needs to extract the transmitted information from the superposition of signals from each path. Due to the different path lengths, such signals do not add up coherently, which may result in fading of the overall received signal.

As the use cases for wireless communication networks continuously evolve, a high mobility scenario is gradually becoming more prominent. For example, high speed trains, self-driving cars, and flying taxis have the potential to travel at speeds of several hundreds of kilometers per hour with passengers requiring high data rates.

We define a *high-mobility wireless channel* as a channel where transmitters, receivers, and many scatterers are moving at different speeds in different directions. For each path, mobility causes different Doppler shifts of the carrier frequency f_c used in the transmitted signal. A Doppler shift by f_d [Hz] is equivalent to modulating the transmitted signal with $e^{j2\pi f_d t}$. The challenge of communicating over a channel like the one illustrated in Fig. 1.2 is that the transmitted signal traveling over different paths is affected by multiple Doppler shifts and delays. Due to the Doppler shifts, the receiver will see a superposition of nonlinearly distorted versions of the transmit signal.

1.2 Waveforms for high-mobility wireless channels

At an abstract level, in a point-to-point communication system, a stream of information symbols $\{a_n\}$ from an alphabet $\mathbb{A}$ is multiplexed in time

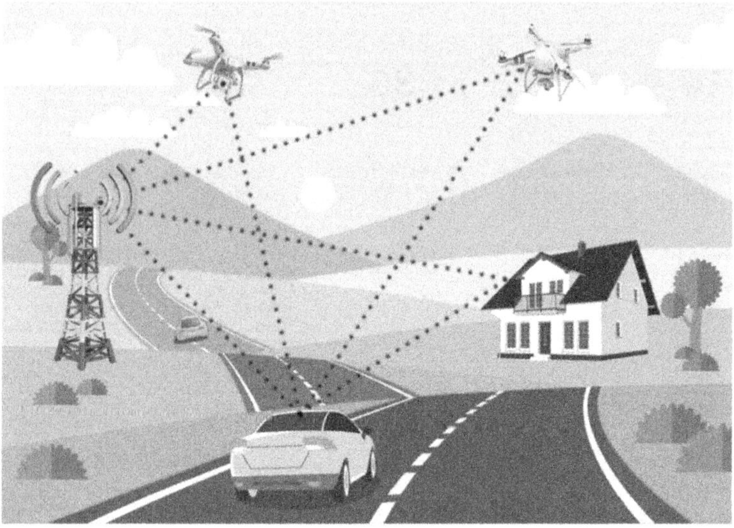

FIGURE 1.2 An example of a high-mobility environment.

and/or frequency. The transmitted signal is formed as

$$s(t) = \sum_n a_n \psi_n(t) \tag{1.1}$$

using a set of orthonormal signals $\Psi_{\mathrm{TX}} = \{\psi_n(t)\}$ such that

$$\langle \psi_n(t), \psi_m(t) \rangle = \int \psi_n(t) \psi_m^*(t)\, dt = \delta_{m,n} = \begin{cases} 1 & m = n, \\ 0 & m \neq n, \end{cases} \tag{1.2}$$

where the operator $(\cdot)^*$ denotes complex conjugation. In the following, for simplicity of exposition, we will neglect the noise term typically appearing in the system.

Assuming $s(t)$ is transmitted over an *ideal* linear channel that introduces no distortion, we receive $r(t) = g_0 s(t - t_0)$, which is a scaled by g_0 and delayed by t_0 version of $s(t)$. In the following, we will assume, without loss of generality, $g_0 = 1$ and $t_0 = 0$, so that we have $r(t) = s(t)$. Due to the linearity of the channel, all the basis signals $\psi_n(t)$ are also received unchanged. We will denote the set of received basis signals as Ψ_{RX}.

An information symbol a_{n_0} can be detected by simply projecting the received signal $s(t)$ onto the same basis signal $\psi_{n_0}(t)$ used to multiplex a_{n_0}, i.e.,

$$\langle s(t), \psi_{n_0}(t) \rangle = \sum_n a_n \langle \psi_n(t), \psi_{n_0}(t) \rangle = a_{n_0}. \tag{1.3}$$

This works thanks to the linearity of the channel and the orthogonality of the basis functions in $\Psi_{RX} = \Psi_{TX}$.

Next, let us assume $s(t)$ is transmitted over a *linear time-invariant* channel characterized by an impulse response $h(t)$. Now, every transmitted basis function is distorted by the convolution with $h(t)$ yielding $\phi_n(t) = h(t) * \psi_n(t)$. Thanks to the linearity of the channel the received signal is given by

$$r(t) = \sum_n a_n \phi_n(t). \tag{1.4}$$

The received basis $\Psi_{RX} = \{\phi_n(t)\}$ is no longer orthonormal and the projection of $r(t)$ on $\psi_{n_0}(t)$ does not yield the symbol a_{n_0}. Instead we have

$$\begin{aligned} r_{n_0} &= \langle s(t), \psi_{n_0}(t)\rangle = a_{n_0}\langle \phi_{n_0}(t), \psi_{n_0}(t)\rangle + \sum_{n\neq n_0} a_n \langle \phi_n(t), \psi_{n_0}(t)\rangle \\ &= a_{n_0}\cos\left(\angle(\phi_{n_0}(t), \psi_{n_0}(t))\right) + \underbrace{\sum_{n\neq n_0} a_n \langle \phi_n(t), \psi_{n_0}(t)\rangle}_{\text{ISI}}, \end{aligned} \tag{1.5}$$

indicating that the sample r_{n_0} contains the useful term a_{n_0}, scaled by the cosine of the angle between $\phi_{n_0}(t)$ and $\psi_{n_0}(t)$. In addition, r_{n_0} contains other interfering terms involving all symbols a_n. The sum of the interfering terms is known as *intersymbol interference* (ISI).

In the presence of ISI, a digital receiver will require a channel equalizer to recover the information symbols. In the case of a channel introducing a very mild distortion, $r_{n_0} \approx a_{n_0}$, which implies that the scalar products among elements of Ψ_{TX} and Ψ_{RX} in the ISI terms are almost zero for any pair n_0 and n ($n \neq n_0$), while $\langle \phi_{n_0}(t), \psi_{n_0}(t)\rangle \approx 1$. Then ISI can be treated as small additive noise, and *symbol-by-symbol detection* can be applied to recover a_{n_0} from r_{n_0}.

When the channel introduces severe distortions, the ISI term is large and may involve many symbols a_n for $n \neq n_0$, which cannot be neglected or assimilated to a small additional noise. In this case, more complex equalization techniques can be applied, such as *maximum likelihood sequence estimation*, which runs through the set of all possible interfering symbols to pick the most likely transmitted a_{n_0}. Unfortunately, the complexity of such an equalizer increases exponentially with the number of symbols a_n involved in the ISI term.

We say the two bases Ψ_{TX} and Ψ_{RX} are *biorthogonal* if

$$\langle \phi_n(t), \psi_m(t)\rangle = \delta_{n,m}. \tag{1.6}$$

As a result, the ISI term at the receiver vanishes. We say Ψ_{TX} and Ψ_{RX} are *quasi-biorthogonal* when the ISI term is much smaller than 1, i.e., when symbol-by-symbol detection is sufficient. In general, we can think that the role of an equalizer at the receiver is to restore biorthogonality (or at least the quasi-biorthogonality) of Ψ_{TX} and Ψ_{RX}.

Although we have discussed *time domain* basis signals, it is important to remember that thanks to Parseval's identity, the orthonormality property of Ψ_{TX} can be equivalently observed with frequency domain signals given by the Fourier transforms $\hat{\psi}_n(f) = \mathcal{F}\{\psi_n(t)\}$ for all n. For the same reason, the ISI term and the cosine term remain unchanged, even if we operate with the *frequency domain* signals (as in the case of OFDM-based systems), i.e.,

$$r_{n_0} = a_{n_0} \cos\left(\angle \hat{\phi}_{n_0}(f), \hat{\psi}_{n_0}(f)\right) + \underbrace{\sum_{n \neq n_0} a_n \langle \hat{\phi}_n(f), \hat{\psi}_{n_0}(f) \rangle}_{\text{ISI}}. \tag{1.7}$$

The above property is not limited to time or frequency domain, but applies to any unitary transformation of the basis signals.

In orthogonal time-frequency space (OTFS), we will consider an orthonormal basis $\tilde{\Psi}_{\text{TX}} = \{\tilde{\psi}_{m,n}(\tau, \nu)\}$ composed of two-dimensional signals in the (τ, ν) *delay-Doppler domain* indexed by m and n. Such basis signals are the Zak transform of some special time domain signals $\Psi_{\text{TX}} = \{\psi_{m,n}(t)\}$ indexed by m (delay) and n (Doppler). As we will see in Chapter 5, the Zak transform is a unitary transformation that guarantees that the time domain basis signals also form an orthonormal basis. In time domain, the transmitter multiplexes information symbols $a_{m,n}$ (arranged in a matrix) as

$$s(t) = \sum_m \sum_n a_{m,n} \psi_{m,n}(t). \tag{1.8}$$

It is interesting to note that, given any received basis Ψ_{RX}, not necessarily orthonormal, it is possible to construct another basis $\Psi^{\perp}$, called the *dual basis*, such that Ψ_{RX} and $\Psi^{\perp}$ form a biorthogonal pair. A receiver using $\Psi^{\perp}$ is based on the well-known *zero-forcing (ZF) equalizer*. Unfortunately, the ZF equalizer also distorts the noise term and degrades the quality of the detection when Ψ_{RX} is not orthogonal. To partially mitigate this problem, a *minimum mean square error (MMSE) equalizer* can be applied to maximize the signal-to-interference-plus-noise ratio on the decisions.

When the channel is known at the transmitter, then precoding can be applied to Ψ_{TX} at the transmitter to guarantee that Ψ_{RX} is orthogonal or quasi-orthogonal, thus simplifying the receiver detection.

Up to now, we have not specified how to select the transmit basis functions. Two design constraints need to be satisfied for any practical communication system: the channel bandwidth and the maximum latency (i.e., the time required for a transmitted information symbol to be detected at the receiver). These two constraints restrict the choice to signals of finite duration that are approximately bandlimited.

A further refinement in the choice of the transmit basis signals is driven by how they interact with the channel. In particular, consider a channel that transforms a transmit basis signal $\psi_n(t)$ into the received basis signal $\phi_n(t)$, where

$$\phi_n(t) = \sum_{i=1}^{P} h_i \psi_{k_i}(t), \tag{1.9}$$

i.e., a weighted sum of a limited number P of other basis functions. In this case, only P terms are involved in the ISI and simpler equalization may be possible. We say that such channel is *sparse* if P is much smaller than the dimension of the basis Ψ_{TX}, since it causes a limited perturbation (loss of orthogonality) of the transmit basis signals Ψ_{TX}.

In OTFS, the transmit basis is the one that provides the simplest representation of a high-mobility multipath channel with P paths. As we will see in Chapter 2, the delay-Doppler domain enables to represent the delay-Doppler channel response of high-mobility multipath channels with P paths as a sparse two-dimensional signal with only P terms,

$$h(\tau, \nu) = \sum_{i=1}^{P} h_i \tilde{\psi}_{l_i,k_i}(\tau, \nu), \tag{1.10}$$

where l_i and k_i are the delay index and the Doppler shift index of path i, for $i = 1, \dots, P$. The use of this basis minimizes the number of terms in the ISI and enables the use of relatively simple iterative detection techniques, as we will see in Chapter 4.

Example

Let us consider a simple example of baseband communications, where information symbols (e.g., ± 1) can be multiplexed in successive time slots of fixed duration T. The set of signals $\Psi_{\text{TX}} = \{p(t - nT)\}_{n=-\infty}^{+\infty}$, where $p(t)$ is a unit-energy pulse signal, zero outside the interval $[0, T)$, forms an orthogonal basis in the time domain, i.e.,

$$\langle p(t - nT), p(t - mT) \rangle = \int_{-\infty}^{+\infty} p(t - nT) p(t - mT)\, dt = \delta_{n,m}. \tag{1.11}$$

The transmitted signal carrying the information symbols a_n is given by

$$s(t) = \sum_{n=-\infty}^{+\infty} a_n p(t - nT). \tag{1.12}$$

When $s(t)$ is transmitted over a bandlimited channel with impulse response $h(t)$, the received signal is given by

$$r(t) = \sum_{n=-\infty}^{+\infty} a_n q(t - nT), \qquad (1.13)$$

where $q(t) = p(t) * h(t)$ is no longer zero outside the interval $[0, T)$, since $h(t)$ is not time limited. The set of signals $\Psi_{\text{RX}} = \{q(t - nT)\}_{n=-\infty}^{+\infty}$ is not an orthonormal basis and is not biorthogonal with Ψ_{TX}, since

$$\langle p(t - nT), q(t - mT) \rangle = \int_{nT}^{(n+1)T} \frac{1}{\sqrt{T}} q(t - mT)\, dt \neq 0 \qquad (1.14)$$

for any $m \neq n$.

Increasing the channel bandwidth or equivalently reducing the transmission rate by increasing T will reduce the effect of ISI, since $q(t - nT) \approx p(t - nT)$ and Ψ_{TX} and Ψ_{RX} become quasi-biorthogonal.

Alternatively, by shaping the pulse $p(t)$ it is possible to slightly reduce the ISI without losing on transmission rate. In some cases, pulse shaping may also result in a slight loss of orthogonality among the transmit basis signals. □

Using the ideas presented above, in Fig. 1.3, we give a graphical representation of the operating principle of multicarrier-based systems and OTFS in static multipath and high-mobility multipath channels. Multicarrier modulation schemes based on orthogonal frequency division multiplexing (OFDM) are suitable for static wireless channels. By adding a cyclic prefix (CP-OFDM in Fig. 1.3(a)) of duration greater than the largest delay path, the distortion introduced by the static multipath channel can be easily compensated at the receiver with symbol-by-symbol detection (single tap equalizer). Although the received signals preserve orthogonality, they will have different powers as a result of a noncoherent combination of different paths. In the case of high mobility, CP-OFDM systems experience degradation even when small Doppler shifts affect the different paths. The resulting received basis signals are no longer orthogonal and will have different powers.

It is interesting to note that the variation in the power of each received basis function has led to the common term "fading channel." However, this misrepresents the fact that the channel itself is not actually fading: the channel is only a collection of scatters whose geometry and mobility generate multiple propagation paths between the transmitter and the receiver.

Even though OFDM offers the advantage of single tap equalization, the nonuniform channel gain on each subcarrier adversely impacts the error performance, since the subcarrier with the smallest channel gain will dominate the overall performance. Moreover, the advantage of single tap equalization is lost in time-varying channels introducing interchannel interference (ICI) due to channel Doppler shifts. These cause large fluctuations in the received power of each subcarrier. A simple solution to the above issue is to increase the overall bandwidth so that the maximum Doppler shift is a very small fraction of the subcarrier spacing. However, this comes at the cost of reduced spectral efficiency.

This approach led to a body of work under the broad term pulse-shaped OFDM (PS-OFDM), focusing on time-frequency pulse shape designs to combat ICI at the receiver. The pulse shape design can be optimized for different criteria such as mitigating ICI, maximizing the received power, improving the spectral efficiency, and reducing the peak-to-average power ratio (PAPR) and out-of-band (OOB) emission. Prior knowledge of channel statistics is crucial in designing these pulse shapes related to minimizing the effects of channel dispersion.

In PS-OFDM in Fig. 1.3(b), we relax the orthogonality constraint of the transmit basis, with the aim of reducing the nonorthogonality of the received basis and their variations in power. However, in some embodiment of PS-OFDM, this is not sufficient to prevent large variations in the power of the different basis signals at the receiver (as shown in Fig. 1.3(b)). Another approach to reducing the received power variability across basis functions is based on precoding information across multiple subcarriers and time slots. This provides large diversity gains but comes at the cost of more complex detection methods.

In OTFS (Fig. 1.3(c)), we use transmit basis functions that are orthogonal in the delay-Doppler domain. With OTFS, the high-mobility (or static) multipath channel marginally reduces the orthogonality of the received basis, and, most importantly, all basis signals are received with the same power.

In this book, we will present how delay-Doppler-based communication systems, such as OTFS, enable to efficiently operate on high-mobility wireless channels. Chapter 2 introduces the high-mobility multipath channel model. Chapter 3 presents the basics of OFDM and its limitations in high-mobility communications. The interaction of OTFS signals with the high-mobility channel will be presented in Chapter 4, and its relation to the Zak transform in Chapter 5. After these two chapters it will become apparent that the received basis in OTFS is no longer orthonormal, and more complex equalization than symbol-by-symbol detection is needed to recover the information symbols. Chapter 6 is dedicated to the detection methods for OTFS and discusses solutions of variable complexity. Chapter 7 discusses channel estimation methods in both the delay-Doppler domain and the delay-time domain. Channel estimation in OTFS offers major savings on pilot overhead given the small number of parameters needed to represent the channel. Chapter 8 presents the extension of OTFS to MIMO

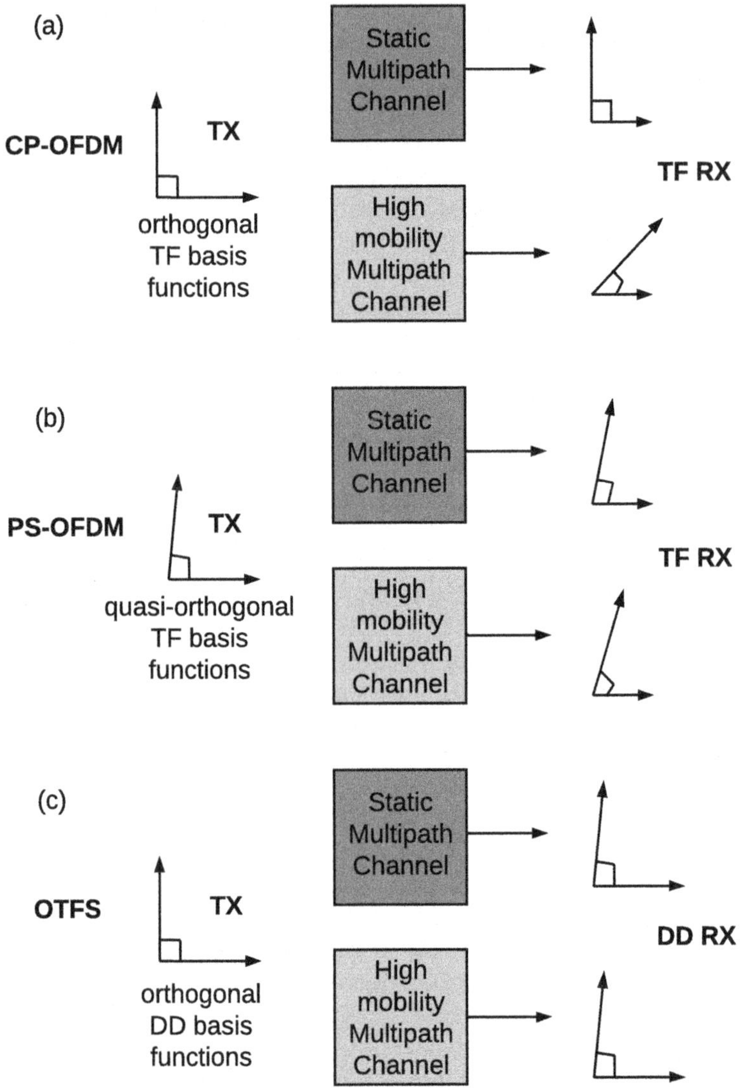

FIGURE 1.3 How the transmit basis functions Ψ_{TX} used to multiplex information symbols are transformed to the receiver basis functions Ψ_{RX} by static multipath and high-mobility channels for (a) CP-OFDM, (b) PS-OFDM, and (c) OTFS. TF: time-frequency, DD: delay-Doppler.

and multiuser uplink and downlink systems. In Chapter 9 we conclude with an outlook at future developments in delay-Doppler communications. Finally, MATLAB® code samples are provided in the Appendix to demonstrate the full implementation of an OTFS system.

1.3 Bibliographical notes

Fundamentals of digital communications can be found in [1,2]. A brief history of wireless communication systems and cellular mobile can be found in [3] and some of the key technologies are discussed in [4–8]. The general theory to model wireless channels can be found in [9–11]. High mobility time-varying channels and their delay-Doppler representations are specifically discussed in [12]. Theory and practice of OFDM techniques for wireless communications are presented in [13]. OTFS modulation was first presented by Hadani *et al.* at the 2017 IEEE Wireless Communications and Networking Conference [14] and its input–output relation was first presented in [15]. Readers can refer to [16–20] for a background on pulse shaping designs for OFDM, based on various optimization criteria. For analysis of multicarrier modulation schemes proposed as improvements over OFDM such as FBMC, UFMC, and GFDM see [21–28].

References

[1] J.G. Proakis, M. Salehi, Digital Communications, fifth edition, McGraw-Hill, 2008.
[2] S. Benedetto, E. Biglieri, Principles of Digital Transmission with Wireless Applications, Springer US, 2002.
[3] S.K. Wilson, S.G. Wilson, E. Biglieri, Academic Press Library in Mobile and Wireless Communications: Transmission Techniques for Digital Communications, Academic Press, Elsevier Ltd, 2016.
[4] G. Stuber, Principles of Mobile Communication, Kluwer Academic Publishers, 2001.
[5] A.J. Viterbi, CDMA: Principles of Spread Spectrum Communication, Addison Wesley, 1995.
[6] A. Paulraj, N. Naar, D. Gore, Introduction to Space-Time Wireless Communications, Cambridge University Press, 2003.
[7] T. Rappaport, R. Heath, R. Daniels, J. Murdock, Millimeter Wave Wireless Communication, Prentice Hall, 2014.
[8] J.G. Andrews, S. Buzzi, W. Choi, S.V. Hanly, A. Lozano, A.C.K. Soong, J.C. Zhang, What will 5G be?, IEEE Journal on Selected Areas in Communications 32 (6) (2014) 1065–1082, https://doi.org/10.1109/JSAC.2014.2328098.
[9] A. Goldsmith, Wireless Communications, Cambridge University Press, 2005.
[10] D. Tse, P. Viswanath, Fundamentals of Wireless Communication, 3rd edition, Cambridge University Press, 2005.
[11] A.F. Molisch, Wireless Communications, second edition, John Wiley & Sons, 2011.
[12] F. Hlawatsch, G. Matz, Wireless Communications over Rapidly Time-Varying Channels, 1st edition, Academic Press, Inc., USA, 2011.
[13] Y.G. Li, G.L. Stuber, Orthogonal Frequency Division Multiplexing for Wireless Communications, Springer, 2006.
[14] R. Hadani, S. Rakib, M. Tsatsanis, A. Monk, A. Goldsmith, A. Molisch, R. Calderbank, Orthogonal time frequency space modulation, in: 2017 IEEE Wireless Communications and Networking Conference (WCNC'17), 2017.
[15] P. Raviteja, K.T. Phan, Y. Hong, E. Viterbo, Interference cancellation and iterative detection for orthogonal time frequency space modulation, IEEE Transactions on Wireless Communications 17 (10) (2018) 6501–6515, https://doi.org/10.1109/TWC.2018.2860011.

[16] W. Kozek, A. Molisch, Nonorthogonal pulseshapes for multicarrier communications in doubly dispersive channels, IEEE Journal on Selected Areas in Communications 16 (8) (1998) 1579–1589, https://doi.org/10.1109/49.730463.
[17] S. Das, P. Schniter, Max-SINR ISI/ICI-shaping multicarrier communication over the doubly dispersive channel, IEEE Transactions on Signal Processing 55 (12) (2007) 5782–5795, https://doi.org/10.1109/TSP.2007.901660.
[18] D. Schafhuber, G. Matz, F. Hlawatsch, Pulse-shaping OFDM/BFDM systems for time-varying channels: ISI/ICI analysis, optimal pulse design, and efficient implementation, in: The 13th IEEE International Symposium on Personal, Indoor and Mobile Radio Communications, 2002, pp. 1–6.
[19] T. Strohmer, S. Beaver, Optimal OFDM design for time-frequency dispersive channels, IEEE Transactions on Communications 51 (7) (2003) 1111–1122, https://doi.org/10.1109/TCOMM.2003.814200.
[20] H. Bölcskei, Orthogonal frequency division multiplexing based on offset QAM, in: Advances in Gabor Analysis, Springer, 2003, pp. 321–352.
[21] M.G. Bellanger, Specification and design of a prototype filter for filter bank based multicarrier transmission, in: 2001 IEEE International Conference on Acoustics, Speech, and Signal Processing. Proceedings, 2001, pp. 1–6.
[22] P. Siohan, C. Siclet, N. Lacaille, Analysis and design of OFDM/OQAM systems based on filterbank theory, IEEE Transactions on Signal Processing 50 (5) (2002) 1170–1183, https://doi.org/10.1109/78.995073.
[23] B. Farhang-Boroujeny, OFDM versus filter bank multicarrier, IEEE Signal Processing Magazine 28 (3) (2011) 92–112, https://doi.org/10.1109/MSP.2011.940267.
[24] V. Vakilian, T. Wild, F. Schaich, S. ten Brink, J.-F. Frigon, Universal-filtered multi-carrier technique for wireless systems beyond LTE, in: 2013 IEEE Globecom Workshops (GC Wkshps), 2013, pp. 223–228.
[25] G. Fettweis, M. Krondorf, S. Bittner, GFDM – generalized frequency division multiplexing, in: VTC Spring 2009 – IEEE 69th Vehicular Technology Conference, 2009, pp. 1–6.
[26] N. Michailow, M. Matthé, I.S. Gaspar, A.N. Caldevilla, L.L. Mendes, A. Festag, G. Fettweis, Generalized frequency division multiplexing for 5th generation cellular networks, IEEE Transactions on Communications 62 (9) (2014) 3045–3061, https://doi.org/10.1109/TCOMM.2014.2345566.
[27] F. Schaich, T. Wild, Waveform contenders for 5G – OFDM vs. FBMC vs. UFMC, in: 2014 6th International Symposium on Communications, Control and Signal Processing (ISCCSP), 2014, pp. 1–6.
[28] M. Matthe, Waveform Design for Generalized Frequency Division Multiplexing: A Survey on Pulse Shaping Filters, AV Akademikerverlag, 2014.

CHAPTER

2

High-mobility wireless channels

OUTLINE

Chapter points

- Wireless channel models for high mobility environments: multipath and Doppler shifts.
- Three domains to represent the channels: frequency-time, delay-time, and delay-Doppler.
- Statistical channel models.

To know what you know and what you do not know, that is true knowledge.
Confucius

Any transmitted signal undergoes variations as it propagates through the wireless channel. Several factors, including channel propagation prop-

https://doi.org/10.1016/B978-0-32-385028-5.00010-4

erties and obstacles in a propagation path, can cause fluctuations in received signal strength, commonly termed fading. Fading can be broadly classified as large-scale fading and small-scale fading.

Large-scale fading involves variations of the average received signal strength due to signal propagation over long distances (over several hundred meters) and a complete or partial line of sight (LoS) path loss due to shadowing caused by the presence of large obstacles in the propagation path.

Small-scale fading, on the other hand, refers to rapid fluctuations that occur within a short period (of the order of seconds) or over short distances (of the order of meters). These fluctuations arise due to constructive and destructive superposition of the transmitted signal echoes that arrive through different propagation paths, a phenomenon known as multipath propagation.

This chapter will focus on small-scale fading, since the physical layer design of a communications system is based on combating the small-scale fading of a wireless channel. As we are concerned with delay-Doppler communications, it is essential to have an in-depth understanding of the delay-Doppler domain representation of a channel before diving into the more practical aspects of transceiver design in later chapters. The delay-Doppler domain has close resemblance with the physical geometric parameters of the environment such as the distance and relative velocity of the reflectors. Under the practical assumption that there are a limited number of reflectors in the vicinity of the receiver, the delay-Doppler domain offers a more compact representation of the geometric channel, as compared to the traditional delay-time or frequency-time domains.

2.1 Input–output model of the wireless channel

Consider the transmission of a signal over a wireless channel. At the transmitter, a baseband signal $s(t)$ of bandwidth B is up-converted to a passband $[f_c - B/2, f_c + B/2]$, where f_c is the carrier frequency used for transmission. At the receiver, the received signal is down-converted to the baseband equivalent signal, denoted by $r(t)$. Since most of the receiver processing, such as demodulation, decoding, and detection, occurs at baseband, we will focus only on the baseband equivalent representation of wireless channels. With the final aim of arriving at the delay-Doppler representation of the channel, we start with the geometric model of a wireless channel.

2.1.1 Geometric model

Geometric models are based on the ray-tracing technique, which is an easy and useful tool to understand how the transmitted waves interfere with the physical channel. In order to derive a deterministic model of the wireless channel, we start with the ray-tracing technique and use the knowledge of the physical geometry of the propagation environment. Due to multipath propagation, the received signal $r(t)$ is an aggregation of delayed, Doppler shifted, and attenuated copies of the transmitted signal $s(t)$. The delay is a function of the length of each propagation path, whereas Doppler shift occurs due to the relative motion in the scene of transmitter, receiver, and reflectors.

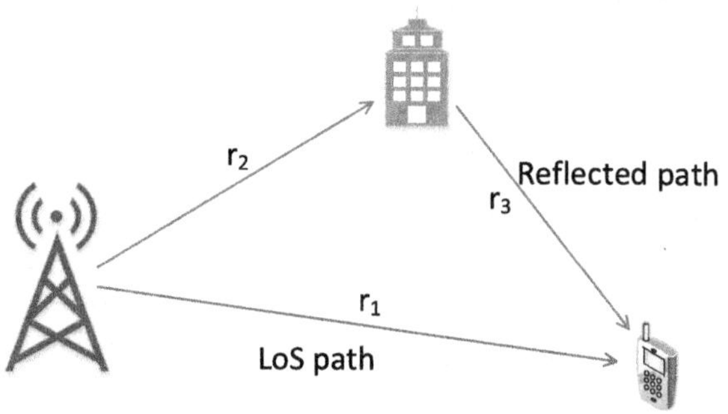

FIGURE 2.1 Paths with different propagation delays.

Let us first consider a simple wireless channel shown in Fig. 2.1, where the transmitter (base-station), the receiver (mobile), and the reflector (building) are static. Since there is no relative motion in the scene, the transmitted signal does not undergo any Doppler shift. However, the difference in the propagation delay of the direct and reflected paths causes two copies of $s(t)$ to arrive at the mobile receiver at different times. The direct path from the base-station to the mobile incurs a propagation delay due to distance r_1. On the other hand, the path reflected from the building has to travel a combined distance of $r_2 + r_3$. Assume that the direct and reflected paths have a baseband equivalent complex gain (attenuation) of g_1 and g_2, respectively. Using the superposition principle, the received signal $r(t)$ can be expressed as

$$r(t) = g_1 s(t - \tau_1) + g_2 s(t - \tau_2), \tag{2.1}$$

where $\tau_1 = r_1/c$ is the delay of the LoS path, $\tau_2 = (r_2 + r_3)/c$ is the delay of the reflected path, and $c = 3 \cdot 10^8$ m/s is the speed of light. The difference in propagation delays $\tau_2 - \tau_1$ is known as the *delay spread.*

In the case of a channel with more than two paths, the delay spread is defined as the difference between the propagation delays of the longest and shortest paths, i.e., $\tau_{max} - \tau_{min}$.

Now consider the case in Fig. 2.2, where the mobile receiver is in a car which is moving towards the base-station with a relative velocity v. We

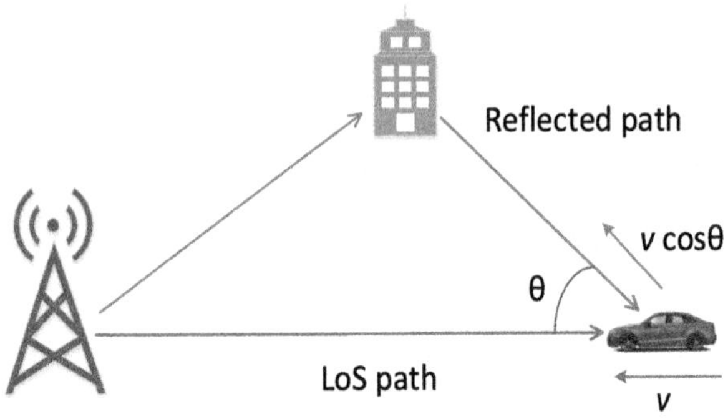

FIGURE 2.2 Paths with different Doppler shifts due to the different angles of arrival.

assume that the bandwidth B of $s(t)$ is very small compared to the carrier frequency f_c, i.e., $f_c \gg B$. The Doppler shift due to a relative velocity υ is given by $\frac{\upsilon}{c} f_c$. The received signal can then be expressed as the sum of delayed and Doppler shifted copies of the transmitted signal as

$$r(t) = \underbrace{g_1 e^{j2\pi \nu_1 (t-\tau_1)}}_{g(\tau_1,t)} s(t-\tau_1) + \underbrace{g_2 e^{j2\pi \nu_2 (t-\tau_2)}}_{g(\tau_2,t)} s(t-\tau_2), \tag{2.2}$$

where $\nu_1 = \frac{\upsilon}{c} f_c$ is the Doppler shift of the LoS path, $\nu_2 = \frac{\upsilon \cos\theta}{c} f_c$ is the Doppler shift of the reflected path, $|\nu_2 - \nu_1|$ is the *Doppler spread*, and the time-dependent functions

$$g(\tau_i, t) = g_i e^{j2\pi \nu_i (t-\tau_i)}, \quad i = 1, 2, \tag{2.3}$$

represent the time-varying attenuation of the propagation paths incurred by delay and Doppler shifts. The delay spread ($\tau_{\max}$) and Doppler spread ($\nu_{\max}$) values for some typical wireless channels are listed in Table 2.1.

In general, the multipath fading channel in (2.3) can then be modeled as an LTI system of the form

$$\boxed{r(t) = \int_0^\infty g(\tau, t) s(t-\tau) d\tau,} \tag{2.4}$$

where $g(\tau, t)$ is the delay-time impulse response of the channel and $0 \leq \tau < \infty$ represents the propagation delay.

The time-frequency impulse response of the channel at a fixed time t can be obtained by taking a Fourier transform along the delay dimension of $g(\tau, t)$,

$$\boxed{H(f, t) = \int_\tau g(\tau, t) e^{-j2\pi f \tau} d\tau.} \tag{2.5}$$

TABLE 2.1 Delay spread (τ_{max}) and Doppler spread (ν_{max}) for some typical wireless channels.

Δr_{max}	Indoor (3 m)	Outdoor (3 km)
τ_{max}	10 ns	10 μs

υ_{max}	$f_c = 2$ GHz	$f_c = 60$ GHz
$\upsilon = 1.5$ m/s $= 5.5$ km/h	$\nu_{max} = 10$ Hz	$\nu_{max} = 300$ Hz
$\upsilon = 3$ m/s $= 11$ km/h	$\nu_{max} = 20$ Hz	$\nu_{max} = 600$ Hz
$\upsilon = 30$ m/s $= 110$ km/h	$\nu_{max} = 200$ Hz	$\nu_{max} = 6$ kHz
$\upsilon = 150$ m/s $= 550$ km/h	$\nu_{max} = 1$ kHz	$\nu_{max} = 30$ kHz

For the general case, where the channel has P paths, each with the gain g_i, the delay τ_i, and the Doppler shift ν_i, $i = 1, \ldots, P$, substituting $g(\tau_i, t)$ in (2.2) into the above equation yields the frequency response as

$$H(f, t) = \sum_{i=1}^{P} g_i e^{-j2\pi \nu_i \tau_i} e^{-j2\pi(f\tau_i - \nu_i t)}. \tag{2.6}$$

In practice, $H(f, t)$ is assumed to be a slowly time-varying function of t. For the special case of a static channel, i.e., $\nu_i = 0, \forall i$, $H(f, t)$ reduces to the time-independent frequency response $H(f)$.

2.1.2 Delay-Doppler representation

We have shown how the different delayed and Doppler shifted components lead to the received signal $r(t)$ and define a model for the wireless channel response. However, the effect of Doppler shift is not immediately evident in the delay-time response $g(\tau, t)$ that we have obtained in the last section. The effect of a scatterer can be represented using the delay (due to distance) and Doppler shift (due to relative motion) undergone by the transmitted signal $s(t)$ that bounces off it. This enables a linear time-varying wireless channel to be completely characterized by the delay-Doppler parameters of the scatterers in the vicinity of the receiver. Since the delay-Doppler response more closely resembles the physical wireless channel, it is useful to have a delay-Doppler representation of the channel.

In general, the small-scale fading effects of a wireless channel can be represented using a small number of scatterers in the vicinity of the receiver. This means that the wireless channel has a sparse representation in the delay-Doppler domain. In order to illustrate this explicitly, let us consider a wireless channel with P propagation paths with distinct delay and Doppler shift parameters.

The two-path input-output relation given in (2.2) can be generalized for a number P of propagation paths as

$$r(t) = \sum_{i=1}^{P} g_i e^{j2\pi \nu_i (t-\tau_i)} s(t-\tau_i), \tag{2.7}$$

where g_i is the path gain and τ_i and ν_i are the delay and Doppler shift, respectively, associated with the i-th path, $i = 1, \ldots, P$.

We define the delay-Doppler response as

$$h(\tau, \nu) = \sum_{i=1}^{P} g_i e^{-j2\pi \nu_i \tau_i} \delta(\tau - \tau_i)\delta(\nu - \nu_i), \tag{2.8}$$

which is a sparse representation of the wireless channel in the delay-Doppler domain due to the limited number of paths P. Then the received signal $r(t)$ can be written as

$$r(t) = \int\int h(\tau, \nu) e^{j2\pi \nu t} s(t-\tau) d\nu d\tau. \tag{2.9}$$

From (2.8), we see that the channel in the delay-Doppler domain is completely represented by the parameters (g_i, τ_i, ν_i) for $i = 1, \ldots, P$. Note that the term $e^{-j2\pi \nu_i \tau_i}$ is a constant phase shift that may be absorbed into the channel coefficient g_i with a slight abuse of notation. Without loss of generality, in the following we will simply assume:

$$\boxed{h(\tau, \nu) = \sum_{i=1}^{P} g_i \delta(\tau - \tau_i)\delta(\nu - \nu_i).} \tag{2.10}$$

Figs. 2.3(a) and 2.3(b) show a simple delay-Doppler representation of the channel between the base-station (Tx) and a vehicle (Rx) on a highway. Both these pictures were taken 100 ms apart in time. Note that the transmitted signal duration is in general less than 10 ms, which is much shorter than the geometric coherence time, i.e., the duration over which physical geometry of the channel can be considered to be invariant. This allows the channel to have a roughly time-invariant response to an impulse in the delay-Doppler domain as long as the impulse is transmitted within the geometric coherence time. The knowledge of the geometric coherence time is crucial in designing the transmission parameters such as signal duration according to the specific channel parameters such as delay spread and Doppler spread.

As an example of a high mobility scenario, consider Figs. 2.3(a) and 2.3(b), where a base-station transmits to a receiver traveling at nearly

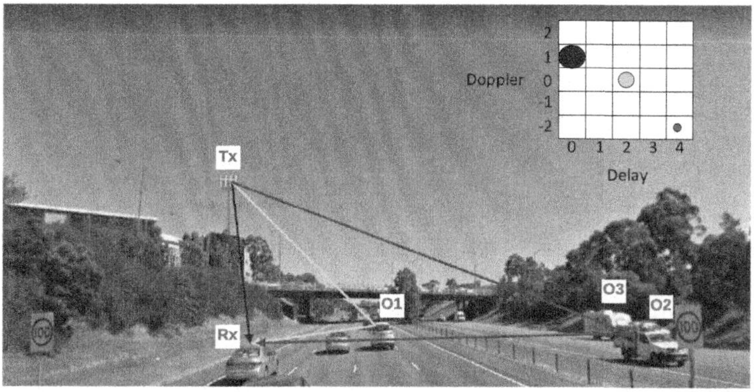

(a) Scenario 1

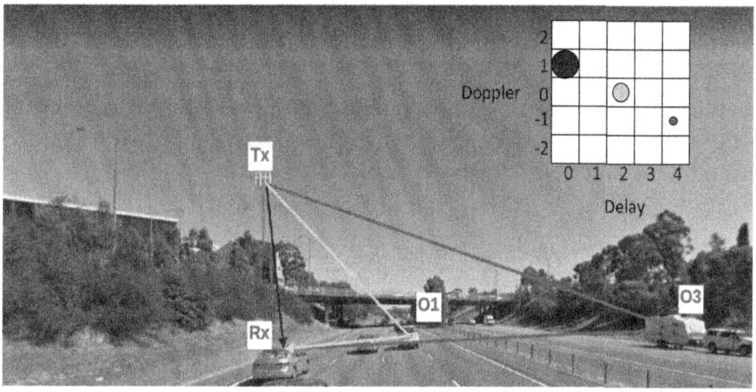

(b) Scenario 2

FIGURE 2.3 An example of high-mobility wireless channel scenarios showing how the delay-Doppler channel response changes when the geometry of the scene changes.

100 km/h in a car. Due to the presence of scatterers, Rx receives an aggregation of P delay and Doppler shifted echoes of the transmitted signal. The scatterers considered in this example are labeled as O1, O2, and O3. The delay-Doppler grid in the figure shows what the mobile user receives if a dot at delay and Doppler position $[0, 0]$ is transmitted. Each colored dot in the delay-Doppler grid corresponds to the propagation path denoted by the same colored ray. The area of the dot in the delay-Doppler grid denotes the gain of each propagation path. Assume that each integer value along the delay and Doppler axis corresponds to a reflector at a distance of 10 m and a speed of 50 km/h relative to Rx. For example, the blue (black in print version) dot corresponds to a reflector traveling at 50 km/h (corresponding to the integer Doppler tap 1) relative to the receiver Rx.

The largest dot in Fig. 2.3(a) corresponds to the LoS path between the base-station and the receiver vehicle Rx, shown by the blue (black in print version) colored ray. It is assigned a delay of 0, being the signal that arrives first, and has a positive Doppler shift, as the car is moving towards the base-station. The yellow (light gray in print version) circle corresponds to the path that is reflected off another vehicle (O1) traveling along the same direction as Rx in the highway, but with lower speed compared to Rx relative to the base-station. The smallest circle corresponds to the longest path due to the reflection from (O2) denoted by the purple (dark gray in print version) dot and rays. The Doppler shift has a negative value as the vehicle is moving away from the base-station and its relative velocity with respect to Rx is zero at this instant due to the relative angles between the vehicles. The reflection from vehicle O3 is ignored due to severe path attenuation as the transmitted wave has to travel a much longer distance than in the case of O2.

Let us now discuss how the channel, and hence the delay-Doppler representation, remains roughly invariant in short intervals of time. As shown in Fig. 2.3(b), in the space of 100 ms, vehicle O2 has moved farther away from Rx and O3 has moved closer to Rx. This causes the reflected wave off O3 (as denoted by the red (mid gray in print version) dot) to be stronger than the one off O2. The Doppler shift due to O3 is less than that due to O2 as O3 is traveling at almost half the speed of O2.

The key conclusion to be drawn from this example is that by aptly designing the frame duration, the delay-Doppler representation of a typical wireless channel can be made to be roughly time-invariant for the duration of a signal frame.

2.2 Continuous-time baseband channel model

In the previous sections, we have looked at general representations of a wireless channel. For a receiver that operates with limited delay and Doppler resolution, it is impossible to observe the true channel parameters. The observed channel is a function of the true channel as well as the delay and Doppler resolution of the receiver. Therefore we start with a continuous-time baseband representation of the channel from the receiver's point of view. Then, we will look at how the receiver sampling results in a discrete-time equivalent baseband channel. In Chapter 4, we will analyze in detail the discrete delay-Doppler channel when a digital receiver is used.

Let the transmitted signal $s(t)$ be of bandwidth $M\Delta f$ [Hz] and duration NT [s]. Consider a baseband equivalent channel model with P propagation paths. For the i-th path, $i = 1, \ldots, P$, the complex path gain is g_i, and the actual delay and Doppler shift are τ_i and ν_i, respectively, given by

$$\tau_i = \frac{\ell_i}{M\Delta f} \leq \tau_{\max} = \frac{\ell_{\max}}{M\Delta f} \quad \nu_i = \frac{\kappa_i}{NT} \text{ with } |\nu_i| \leq \nu_{\max}, \tag{2.11}$$

where $\ell_i, \kappa_i \in \mathbb{R}$ are the *normalized delay* and *normalized Doppler shift*, respectively, and $\ell_{\max} \in \mathbb{R}$ is the *normalized delay* associated with $\tau_{\max}$.

We assume that the channel is *underspread*, i.e., $\tau_{\max}\nu_{\max} \ll 1$ and $T\Delta f = 1$. Under the underspread assumption, we have $\ell_{\max} < M$ and the normalized Doppler shifts $-N/2 < \kappa_i < N/2$. Recalling from the previous sections, since the number of channel coefficients P in the delay-Doppler domain is typically limited, the delay-Doppler channel response has a sparse representation:

$$h(\tau, \nu) = \sum_{i=1}^{P} g_i \delta(\tau - \tau_i)\delta(\nu - \nu_i). \tag{2.12}$$

We let $\mathcal{L} = \{\ell_i\}$ of size $|\mathcal{L}|$ be the set of distinct *normalized delays* among the P paths in the delay-Doppler domain, we let $\mathcal{K}_\ell = \{\kappa_i \mid \ell = \ell_i\}$ be the set of *normalized Doppler shifts* for each path with normalized delay ℓ_i, and we let

$$\nu_\ell(\kappa) = \begin{cases} g_i, & \text{if } \ell = \ell_i \text{ and } \kappa = \kappa_i, \\ 0, & \text{otherwise} \end{cases} \tag{2.13}$$

be the *Doppler response* at delay ℓ. Then we can rewrite (2.12) as

$$h(\tau, \nu) = \sum_{\ell \in \mathcal{L}} \sum_{\kappa \in \mathcal{K}_\ell} \nu_\ell(\kappa)\delta(\tau - \ell T/M)\delta(\nu - \kappa\Delta f/N). \tag{2.14}$$

Note that the continuous delay-time channel response is given by

$$\boxed{g(\tau, t) = \int_\nu h(\tau, \nu) e^{j2\pi\nu(t-\tau)}\, d\nu.} \tag{2.15}$$

Substituting (2.14) into (2.15) yields the corresponding delay-time channel response, for all $\ell \in \mathcal{L}$, as

$$g(\tau, t) = \sum_{\ell \in \mathcal{L}} \sum_{\kappa \in \mathcal{K}_\ell} \nu_\ell(\kappa) e^{j2\pi\kappa\frac{\Delta f}{N}(t - \ell T/M)} \delta(\tau - \ell T/M). \tag{2.16}$$

Evaluating (2.16) at $\tau = \ell T/M$ for all $\ell \in \mathcal{L}$ yields

$$g(\tau = \ell T/M, t) = \sum_{\kappa \in \mathcal{K}_\ell} \nu_\ell(\kappa) e^{j2\pi\kappa\frac{\Delta f}{N}(t - \ell T/M)}. \tag{2.17}$$

In the special case of a static channel with no Doppler spread, i.e., $\kappa = 0$, (2.16) reduces to

$$g(\tau = \ell T/M, t) = \nu_\ell(0). \tag{2.18}$$

2.3 Discrete-time baseband channel model

In the previous section, we have looked at the continuous-time channel model. When modeling a bandpass communication system, it is convenient to work with a discrete baseband equivalent representation of the system. At the transmitter, a signal of bandwidth $B = M\Delta f$ is up-converted to a carrier frequency f_c to occupy a bandpass channel, assuming $f_c \gg B$. At the receiver, the channel impaired signal is down-converted to baseband and sampled at $f_s = B = M\Delta f$ Hz, yielding NM complex samples per frame of duration NT [s].

By sampling the received waveform $r(t)$ at $t = qT/M$, where $q = 0, \ldots, NM-1$, and discretizing delay variable $\tau = lT/M$ for $l = 0, \ldots, M-1$, we have the discrete baseband delay-time channel response in (2.16) as

$$g^{\mathrm{s}}[l,q] = g(lT/M, qT/M) = \sum_{\ell \in \mathcal{L}} \left(\sum_{\kappa \in \mathcal{K}_\ell} \nu_\ell(\kappa) z^{\kappa(q-l)} \right) \mathrm{sinc}(l - \ell), \tag{2.19}$$

where $\mathrm{sinc}(x) = \sin(\pi x)/(\pi x)$ and $z = \mathrm{e}^{\frac{j2\pi}{NM}}$. Thus the discrete delay-time input-output relation is given as

$$r[q] = r(qT/M) = \sum_{l=0}^{M-1} g^{\mathrm{s}}[l,q] s[q-l]. \tag{2.20}$$

Note that due to fractional delays, the sampling at the receiver introduces interference between Doppler responses at different delays. This is due to sinc reconstruction of the delay-time response at fractional delay points ($\ell \in \mathcal{L}$). However, in typical wideband systems, the channel path delays can be approximated to integer multiples of T/M, without loss of accuracy, i.e., $\ell = l \in \mathbb{Z}$, and the sinc function in (2.19) reduces to

$$\mathrm{sinc}(l - \ell) = \begin{cases} 1, & \text{if } \ell = l, \\ 0, & \text{otherwise.} \end{cases} \tag{2.21}$$

Consequently, the relation between the actual Doppler response and the sampled time domain channel at each integer delay tap $l \in \mathcal{L}$ in (2.19) reduces to

$$g^{\mathrm{s}}[l,q] = \begin{cases} \sum_{\kappa \in \mathcal{K}_l} \nu_l(\kappa) z^{\kappa(q-l)} & \text{for } \ell = l \in \mathcal{L}, \\ 0 & \text{otherwise.} \end{cases} \tag{2.22}$$

Here, we remind the reader that the effective channel as seen by the receiver depends on the actual channel response as well as the operation parameters (delay resolution) of the receiver. Further, we denote $l_{\max} = \max(\mathcal{L})$ to be the *maximum channel delay tap*.

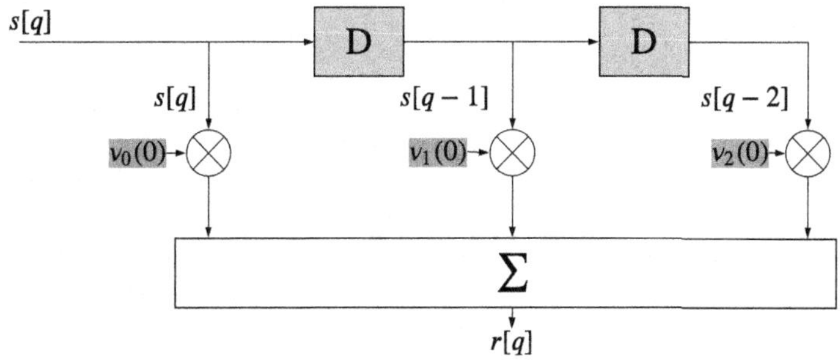

(a) Linear time-invariant channel

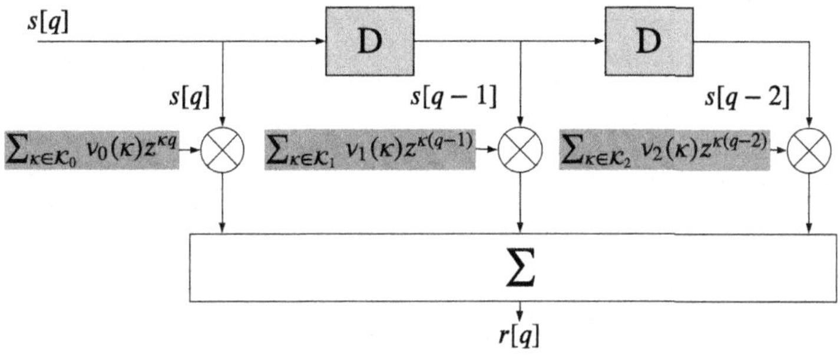

(b) Linear time-varying channel

FIGURE 2.4 The TDL models for static and high-mobility channels with integer delay taps $\mathcal{L} = \{0, 1, 2\}$.

Fig. 2.4 shows the delay-time channel for static and mobile channels, respectively, represented using a tapped delay line (TDL) model for the set of integer delay taps $\mathcal{L} = \{0, 1, 2\}$. For the static case in Fig. 2.4(a), the channel tap corresponding to each delay tap l remains constant since from (2.18) $\nu_l(\kappa) = 0$ for $\kappa \neq 0$. However, for the time-variant channel case in Fig. 2.4(b), each delay tap coefficient is the sum of distinct Doppler paths within the same delay tap, each affected by a time-varying phase rotation.

2.4 Relation among different channel representations

From previous sections, when $s(t)$ is the transmitted time domain signal, the received signal $r(t)$ in the presence of time-variant multipath channel (also known as doubly dispersive channel) can be written as

$$r(t) = \int_{\tau} g(\tau, t) s(t - \tau) d\tau \tag{2.23}$$

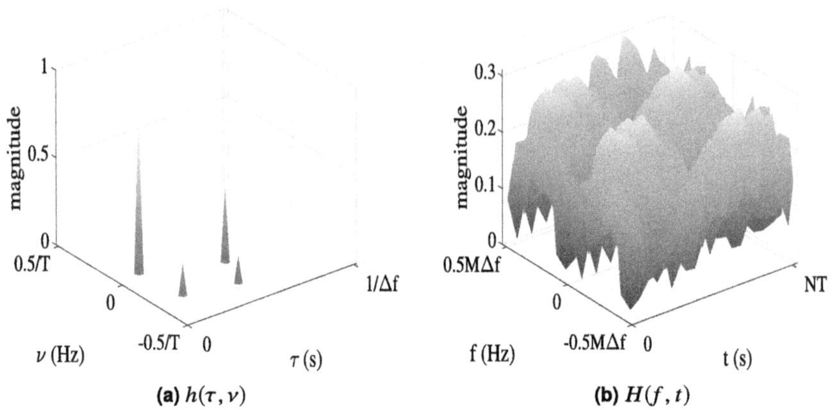

FIGURE 2.5 The continuous delay-Doppler vs time-frequency channel representation of a high-mobility multipath channel (linear time-varying).

$$= \int_f H(f,t)S(f)e^{j2\pi ft}df \tag{2.24}$$

$$= \int_\nu \int_\tau h(\tau,\nu)s(t-\tau)e^{j2\pi\nu t}d\tau d\nu, \tag{2.25}$$

where $S(f)$ is the Fourier transform of $s(t)$.

The three equivalent relations in (2.23)–(2.25) can be interpreted as follows. The channel $g(\tau, t)$ in (2.23) represents the time-varying impulse response and the relation can be seen as a straightforward generalization of the LTI system. The relation in (2.24) describes the time-frequency channel and an OFDM-based system is defined by this relation. Finally, the relation in (2.25) describes the delay-Doppler channel and an OTFS system is based on this relation.

Now, the relation between the time-frequency ($H(f,t)$) and delay–Doppler ($h(\tau,\nu)$) channel responses can be given by a pair of two-dimensional symplectic Fourier transforms (SFTs) as

$$\boxed{h(\tau,\nu) = \text{SFT}\{H(f,t)\} = \iint H(f,t)e^{-j2\pi(\nu t - f\tau)}dt df,} \tag{2.26}$$

$$\boxed{H(f,t) = \text{ISFT}\{h(\tau,\nu)\} = \iint h(\tau,\nu)e^{j2\pi(\nu t - f\tau)}d\tau d\nu,} \tag{2.27}$$

where (2.26) and (2.27) define the SFT and ISFT operations, respectively.

Figs. 2.5 and 2.6 illustrate the sparsity of the delay-Doppler channel response in comparison with the time-frequency channel response of linear time-varying and time-invariant channels. The nonzero coefficients in the

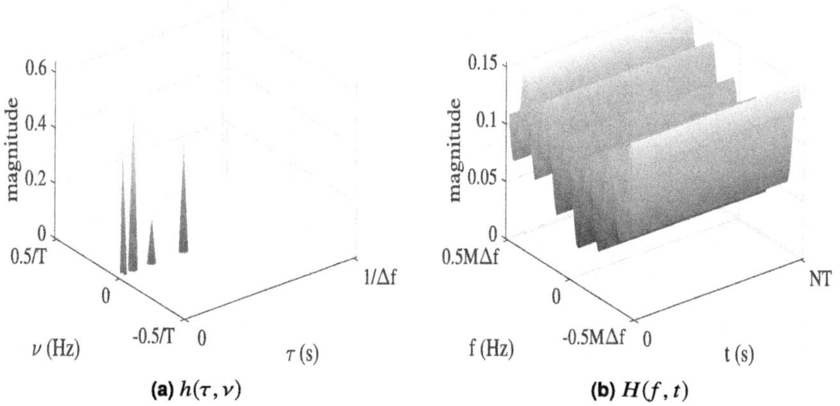

FIGURE 2.6 The continuous delay-Doppler vs. time-frequency channel representation of a static multipath channel (linear time-invariant).

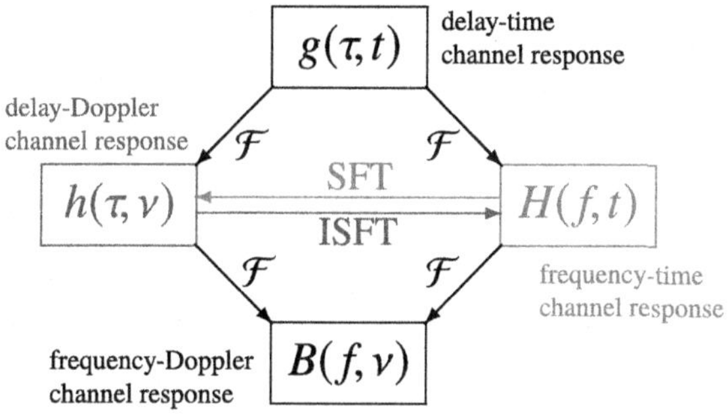

FIGURE 2.7 Different domain representations of a time-variant multipath channel impulse response $g(\tau,t)$, also denoted as the delay-time channel response.

delay-Doppler plane at location (τ, ν) represent the magnitude of the propagation gain of a path with the corresponding delay and Doppler shift. The number of paths in this case is set to $P = 4$. It can be observed that the corresponding frequency-time channel in Fig. 2.5 is time-varying and hence requires more time-frequency coefficients to accurately represent the channel. However, for a static channel, the frequency-time channel is time-invariant with P paths as shown in Fig. 2.6. As seen by the delay-Doppler representation, all the paths have Doppler shift $\nu_i = 0$, where $i = 1, \ldots, P$. Fig. 2.6 represents a frequency selective channel, whereas Fig. 2.5 is both time and frequency selective, or in other words, doubly selective.

In summary, Fig. 2.7 illustrates the relations between the four equivalent time-variant multipath channels defined in the literature, where $\mathcal{F}$

denotes the Fourier operation and $B(f, \nu)$ denotes the Doppler-variant frequency response.

2.5 Channel models for numerical simulations

We adopt the tapped delay line model of the wireless channel shown in Fig. 2.4(b) for our simulations. The channel gain g_i is modeled as a circular symmetric Gaussian complex random variable[1] independent across all paths for $i = 1, \ldots P$, whose envelope is Rayleigh distributed with variance σ_i^2. In general, the channel characteristics can be specified using the power delay profile and maximum Doppler spread of the propagation environment.

2.5.1 Standard wireless mobile multipath propagation scenarios

Here, we list some standard 3GPP multipath fading channel profiles, which will be used to generate channel models for simulations in later chapters. The fading models listed here represent low, medium, and high delay spread environments. The power delay profiles are given in Tables 2.2, 2.3, 2.4.

TABLE 2.2 Extended Pedestrian A (EPA) model.

Excess tap delay (ns)	0	30	70	90	110	190	410
Relative power (dB)	0.0	−1.0	−2.0	−3.0	−8.0	−17.2	−20.8

TABLE 2.3 Extended Vehicular A (EVA) model.

Excess tap delay (ns)	0	30	150	310	370	710	1090	1730	2510
Relative power (dB)	0.0	−1.5	−1.4	−3.6	−0.6	−9.1	−7.0	−12.0	−16.9

TABLE 2.4 Extended Typical Urban (ETU) model.

Excess tap delay (ns)	0	50	120	200	230	500	1600	2300	5000
Relative power (dB)	−1.0	−1.0	−1.0	0.0	0.0	0.0	−3.0	−5.0	−7.0

Let $\nu_{\max} = f_c u_{\max}/c$ be the maximum Doppler shift of the overall multipath channel. We assume a single Doppler shift is associated with the i-th delay path and follows the classic Jakes spectrum, i.e., $\nu_i = \nu_{\max}\cos(\theta_i)$, where θ_i is uniformly distributed over $[-\pi, \pi]$. The MATLAB® code for

[1]The complex channel gain is $g_i = a_i + jb_i$, where a_i and b_i are zero-mean, independent and identically distributed Gaussian random variables with variance $\frac{\sigma_i^2}{2}$, i.e., $a_i, b_i \sim \mathcal{N}(0, \frac{\sigma_i^2}{2})$.

generating the delay-Doppler channel coefficients for a particular UE speed $u_{\max}$ is given in MATLAB code 6 in Appendix C.

2.5.2 Synthetic propagation scenario

For the purpose of studying the effect of multipath on error performance in some of the simulations, we propose a simple channel model with $|\mathcal{L}|$ distinct delay paths, randomly chosen in the interval $[0, \tau_{\max}]$, with equal path gain. Each delay index can have a number of paths with different Doppler shifts uniformly distributed in the interval $[-\nu_{\max}, \nu_{\max}]$. Recall that $\mathcal{K}_\ell$ is the set containing the Doppler shifts of all paths with normalized delay ℓ, where $\ell \in \mathcal{L}$. The total number of paths $P = \sum_{\ell \in \mathcal{L}} |\mathcal{K}_\ell|$. Note that this scenario is unlikely to be found in any physical channel and is purely used to study the receiver performance under a simple set of controlled parameters. See MATLAB code 7 in Appendix C for generating the synthetic channel parameters for arbitrary delay and Doppler spread.

2.6 Bibliographical notes

We assume the reader is familiar with the basic notions of digital communications [1–3]. A more basic understanding of the exposition presented in this chapter can be found in [4–10]. Further details regarding the 3GPP standard channel models are available in [11]. General OFDM techniques for wireless communications were presented in [12]. OTFS modulation was first presented by Hadani et al. at the 2017 IEEE Wireless Communications and Networking Conference [13]. The channel representations of the different domains in various formats are discussed in [14–20].

References

[1] B. Sklar, F.J. Harris, Digital Communications: Fundamentals and Applications, Prentice-Hall, 1988.

[2] S. Benedetto, E. Biglieri, Principles of Digital Transmission with Wireless Applications, Springer US, 2002.

[3] J.G. Proakis, M. Salehi, Digital Communications, McGraw-Hill, 2008.

[4] B. Sklar, Rayleigh fading channels in mobile digital communication systems. I. Characterization, Philips Journal of Research 35 (7) (1997) 90–100.

[5] D. Tse, P. Viswanath, Fundamentals of Wireless Communication, 3rd edition, Cambridge University Press, 2005.

[6] A. Goldsmith, Wireless Communications, Cambridge University Press, 2005.

[7] G. Matz, F. Hlawatsch, Time-varying communication channels: fundamentals, recent developments, and open problems, in: 2006 14th European Signal Processing Conference, 2006, pp. 1–6.

[8] F. Hlawatsch, G. Matz, Wireless Communications over Rapidly Time-Varying Channels, 1st edition, Academic Press, Inc., USA, 2011.

[9] A.F. Molisch, Wireless Communications, second edition, John Wiley & Sons, 2011.

[10] P. Montezuma, F. Silva, R. Dinis, Frequency-Domain Receiver Design for Doubly Selective Channels, Taylor & Francis, 2017.

[11] European Telecommunications Standards Institute.
[12] Y.G. Li, G.L. Stuber, Orthogonal Frequency Division Multiplexing for Wireless Communications, Springer, 2006.
[13] R. Hadani, S. Rakib, M. Tsatsanis, A. Monk, A.J. Goldsmith, A.F. Molisch, R. Calderbank, Orthogonal time frequency space modulation, in: 2017 IEEE Wireless Communications and Networking Conference (WCNC), 2017, pp. 1–6.
[14] P. Raviteja, K.T. Phan, Y. Hong, E. Viterbo, Interference cancellation and iterative detection for orthogonal time frequency space modulation, IEEE Transactions on Wireless Communications 17 (10) (2018) 6501–6515, https://doi.org/10.1109/TWC.2018.2860011.
[15] K.R. Murali, A. Chockalingam, On OTFS modulation for high-Doppler fading channels, in: 2018 Information Theory and Applications Workshop (ITA), 2018, pp. 1–6.
[16] P. Raviteja, Y. Hong, E. Viterbo, E. Biglieri, Practical pulse-shaping waveforms for reduced-cyclic-prefix OTFS, IEEE Transactions on Vehicular Technology 68 (1) (2019) 957–961, https://doi.org/10.1109/TVT.2018.2878891.
[17] A. Farhang, A. RezazadehReyhani, L.E. Doyle, B. Farhang-Boroujeny, Low complexity modem structure for OFDM-based orthogonal time frequency space modulation, IEEE Wireless Communications Letters 7 (3) (2018) 344–347, https://doi.org/10.1109/LWC.2017.2776942.
[18] A. Rezazadehreyhani, A. Farhang, A. Ji, R.R. Chen, B. Farhang-Boroujeny, Analysis of discrete-time MIMO OFDM-based orthogonal time frequency space modulation, in: 2018 IEEE International Conference on Communications, 2018, pp. 1–6.
[19] W. Shen, L. Dai, J. An, P.Z. Fan, R.W. Heath, Channel estimation for orthogonal time frequency space (OTFS) massive MIMO, IEEE Transactions on Signal Processing 67 (16) (2019) 4204–4217, https://doi.org/10.1109/TSP.2019.2919411.
[20] T. Thaj, E. Viterbo, Y. Hong, Orthogonal time sequency multiplexing modulation: analysis and low-complexity receiver design, IEEE Transactions on Wireless Communications 20 (12) (2021) 7842–7855, https://doi.org/10.1109/TWC.2021.3088479.

CHAPTER

3

OFDM review and its limitations

OUTLINE

ChapterPoints

- OFDM modulation and demodulation.
- OFDM advantages and limitations.
- Issues of OFDM in high-mobility channels.

Delay-Doppler Communications
https://doi.org/10.1016/B978-0-32-385028-5.00011-6

3.1 Introduction

Orthogonal frequency division multiplexing (OFDM) is a widely used modulation scheme that forms the basis of 4G/5G mobile communications systems. In the wideband multicarrier scheme, information symbols are multiplexed on closely spaced orthogonal subcarriers. This allows data to be transmitted on parallel channels as long as the orthogonality of subcarriers is not disrupted by the wireless channel. The key advantage of such data transmission is that the orthogonality property enables the use of a single tap equalizer to detect the transmitted data at the receiver. Thus, it provides a low complexity solution to reliable communication in frequency selective channels such as the static multipath wireless channel.

Even though OFDM with cyclic prefix (CP) has been the primary waveform candidate for 4G/5G systems, it suffers from some limitations such as high peak-to-average power ratio (PAPR), out-of-band (OOB) emissions, sensitivity to carrier frequency offsets (CFO), and severe loss of orthogonality in high-mobility wireless channels. Some of these issues can be mitigated by modifying OFDM as, for example, universal filtered multicarrier (UFMC) or filter bank multicarrier (FBMC). These are based on the idea of spreading information symbols across several independent subcarriers. Other schemes like generalized frequency division multiplexing (GFDM) have been developed, where the information symbols are spread across both subcarriers and time slots, using nonorthogonal transformations (filtering).

Even though there are many variations of OFDM, most schemes can be generalized under the term *pulse-shaped* OFDM (PS-OFDM). In this chapter, we first discuss the general OFDM structure with arbitrary pulse shaping. Then our primary focus will be on PS-OFDM with rectangular pulse shaping waveforms, coinciding with the standard *cyclic prefix* OFDM (CP-OFDM).

3.2 OFDM system model

Consider an $M \times N$ OFDM system, where M and N denote the number of subcarriers and time slots, respectively. The OFDM signal has a total bandwidth of $B = M\Delta f$ and occupies a frame duration of $T_{\mathrm{f}} = NT = NMT_s$, where $\Delta f = 1/T$ denotes the subcarrier spacing and $T = MT_s$ denotes an OFDM symbol duration. Here $T_s = 1/f_s$ is the sampling interval and f_s is the sampling frequency.

We consider a static multipath channel, where $\tau_{\max}$ is the maximum delay spread and $l_{\max} < M$ is the maximum channel delay tap. Typically, OFDM adopts a CP of sufficient length $L_{\mathrm{CP}} \geq l_{\max}$ for reliable communications. In this chapter, we choose $L_{\mathrm{CP}} = l_{\max}$.

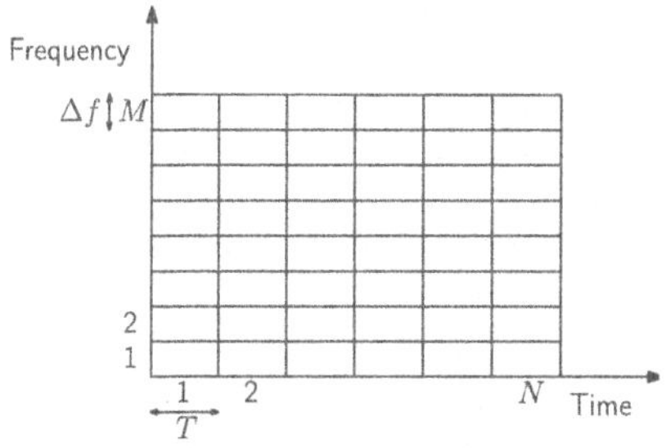

FIGURE 3.1 The discrete time-frequency grid (Λ).

3.2.1 Generalized multicarrier modulation

The discretized *time-frequency domain* with sampling points at multiples of $\Delta f = 1/T$ spaced by T can be represented as an $M \times N$ array of points, shown in Fig. 3.1, as

$$\Lambda = \{(m\Delta f, nT),\ m = 0, \ldots, M-1, n = 0, \ldots, N-1\}$$

for positive integers M, N.

We define the information symbol matrix $\mathbf{X}$ in the time-frequency plane, with elements $\mathbf{X}[m,n]$, $m = 0, \ldots, M-1$, $n = 0, \ldots, N-1$, taken from the QAM alphabet $\mathbb{A} = \{a_1, \cdots, a_Q\}$ of size Q. Each column of $\mathbf{X}$ contains N OFDM symbols.

Explicitly, the transmitted signal of an OFDM-based multicarrier modulation is given by

$$\boxed{s(t) = \sum_{n=0}^{N-1}\sum_{m=0}^{M-1} \mathbf{X}[m,n] g_{\text{tx}}(t-nT) e^{j2\pi m\Delta f(t-nT)},} \tag{3.1}$$

where $g_{\text{tx}}(t)$ for $0 \le t < T$ denotes the pulse shaping waveform used to shape the continuous-time transmitted signal.

As discussed in Chapter 1, the transmitter multiplexes the information symbols using a set of basis signals parametrized by indices m and n:

$$\Psi_{\text{TX}} = \{\psi_{m,n}(t)\} = \{g_{\text{tx}}(t-nT)e^{j2\pi m\Delta f(t-nT)}\}_{0 \le m < M, 0 \le n < N},$$

with $g_{\text{tx}}(t) \ge 0$ for $0 \le t < T$ and zero otherwise. To demultiplex the information, the receiver uses the receive basis signals

$$\Psi_{\text{RX}} = \{\phi_{m,n}(t)\} = \{g_{\text{rx}}(t-nT)e^{j2\pi m\Delta f(t-nT)}\}_{0 \le m < M, 0 \le n < N},$$

with $g_{\text{rx}}(t) \ge 0$ for $0 \le t < T$ and zero otherwise.

Since $T\Delta f = 1$, both bases are orthonormal, i.e., $\langle \psi_{m,n}(t), \psi_{m',n'}(t)\rangle = \delta_{m,m'}\delta_{n,n'}$ and $\langle \phi_{m,n}(t), \phi_{m',n'}(t)\rangle = \delta_{m,m'}\delta_{n,n'}$, where the scalar product of signals $a(t)$ and $b(t)$ is defined as

$$\langle a(t), b(t)\rangle = \int_{-\infty}^{+\infty} a(t)b^*(t)dt,$$

where $b^*(t)$ is the complex conjugate of $b(t)$, and if $b(t)$ is real, then $b^*(t) = b(t)$.

We define the *cross-ambiguity function* between two signals $g_1(t)$ and $g_2(t)$ as

$$A_{g_1,g_2}(f,t) \triangleq \int g_1(t')g_2^*(t'-t)e^{-j2\pi f(t'-t)}dt', \tag{3.2}$$

which defines the correlation between $g_1(t)$ and a version of $g_2(t)$ delayed by t and frequency shifted by f, for any t and f in the time-frequency plane.

Let $r(t)$ be the received time domain signal after $s(t)$ passes through a time-frequency selective wireless channel. The received time-frequency samples are obtained by projecting $r(t)$ on each $\phi_{m,n}(t)$, which equates to computing the cross-ambiguity function $A_{r,g_{\text{rx}}}(f,t)$ and sampling it at the grid points in Λ as

$$\begin{aligned} Y(f,t) &= A_{r,g_{\text{rx}}}(f,t) \triangleq \int r(t)g_{\text{rx}}^*(t'-t)e^{-j2\pi f(t'-t)}dt', \\ \mathbf{Y}[m,n] &= Y(f,t)|_{f=m\Delta f, t=nT}. \end{aligned} \tag{3.3}$$

The cross-ambiguity function of the pulse shaping waveforms $A_{g_{\text{tx}},g_{\text{rx}}}$ determines if Ψ_{TX} and Ψ_{TX} are biorthogonal, i.e., if

$$\langle g_{\text{tx}}(t-n'T)e^{-j2\pi m'\Delta f t}, g_{\text{rx}}(t-nT)e^{-j2\pi m\Delta f t}\rangle = \delta_{n,n'}\delta_{m,m'} \tag{3.4}$$

for $m = 0, \ldots, M-1$, $n = 0, \ldots, N-1$. In such a case, the samples $\mathbf{Y}[m,n]$ would not suffer from intersymbol interference, and the information symbols could be recovered with symbol-by-symbol detection, as discussed in Chapter 1. Biorthogonality would be guaranteed if the cross-ambiguity function was given by

$$A_{g_{\text{tx}},g_{\text{rx}}}(f,t) = \delta(f)\delta(t). \tag{3.5}$$

However, such a cross-ambiguity function is not physically realizable and would correspond to ideal pulse shaping waveforms g_{tx} and g_{rx}, which unfortunately do not exist in practice.

As an example, let us consider practical pulse shaping waveforms $g_{\text{tx}}(t)$ and $g_{\text{rx}}(t)$ to be unit-energy square pulses of duration T, zero outside the

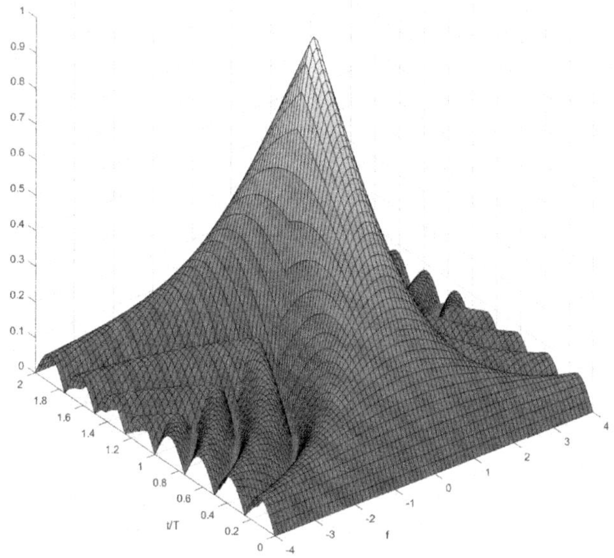

FIGURE 3.2 The magnitude of the cross-ambiguity function of square pulses.

interval $[0, T)$. We have

$$A_{g_{\text{tx}},g_{\text{tx}}}(f,t)=\begin{cases}\frac{j}{2\pi fT}(1-e^{j2\pi|f|t}) & \text{for } 0\le t<T,\\ \frac{j}{2\pi fT}(1-e^{j2\pi|f|(2T-t)}) & \text{for } T\le t<2T.\end{cases} \tag{3.6}$$

The magnitude of the cross-ambiguity function is shown in Fig. 3.2 and its sections along different frequencies and times are shown in Figs. 3.3 and 3.4.

Fig. 3.5 illustrates how the zeros of the cross-ambiguity functions include all the points on the grid Λ (red circles) for any shifts in time by integer multiples of T or in frequency by integer multiples of Δf.

Assuming the channel introduces a delay τ_0 and a Doppler shift ν_0 to the transmitted basis signal $\psi_{m',n'} = g_{\text{tx}}(t-n'T)e^{-j2\pi(m'\Delta f)(t-n'T)}$, the correlation with $\phi_{m,n}(t)$ for $m \neq m'$ and $n \neq n'$ is given by

$$\begin{aligned}&\langle g_{\text{tx}}\big(t-n'T-\tau_0\big)e^{-j2\pi(m'\Delta f+\nu_0)(t-n'T-\tau_0)}, g_{\text{rx}}(t-nT)e^{-j2\pi m\Delta f t}\rangle\\ &\quad = A_{g_{\text{rx}},g_{\text{tx}}}\big((m-m')\Delta f+\nu_0, (n-n')T-\tau_0\big),\end{aligned} \tag{3.7}$$

which is mostly nonzero for arbitrary $0<\tau_0<T$ and $0<\nu_0<\Delta f$.

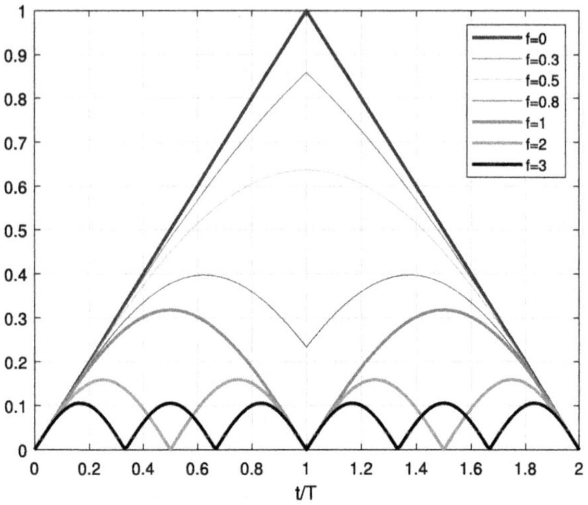

FIGURE 3.3 The magnitude of the cross-ambiguity function of square pulses for different values of f.

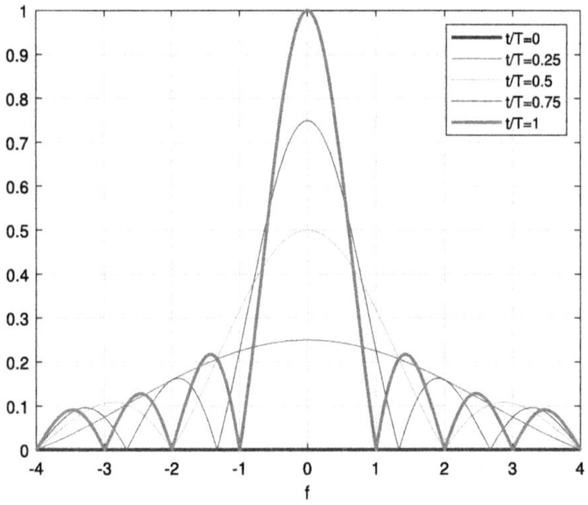

FIGURE 3.4 The magnitude of the cross-ambiguity function of square pulses for different t/T.

This leads to ISI and inevitable performance degradation, particularly when considering high-mobility channels with nonnegligible τ_0 and ν_0. Specific pulse shaping waveforms g_{tx} and g_{rx} can be designed to mitigate in part such performance degradation. Such methods fall under the broad class of PS-OFDM systems.

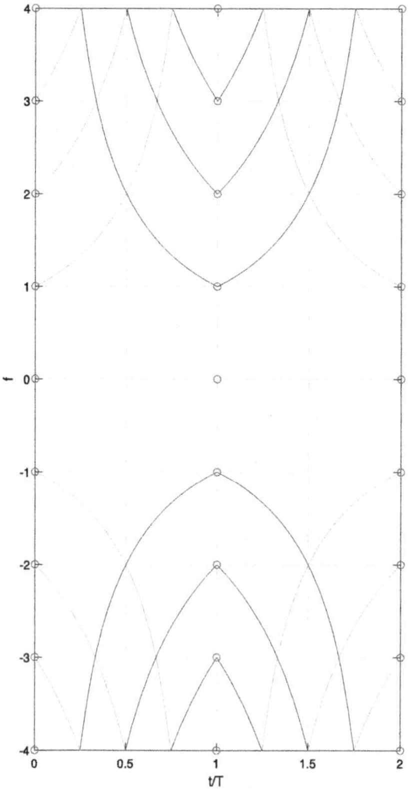

FIGURE 3.5 The zeros of the cross-ambiguity function in the time-frequency plane.

In this chapter, we focus on the PS-OFDM using rectangular pulse shaping waveforms, i.e., the traditional CP-OFDM, over static multipath channels.

3.2.2 OFDM transmitter

Figs. 3.6 and 3.7 show the transmitter and receiver block diagrams, containing NM information symbols, taken from the QAM modulation alphabet $\mathbb{A} = \{a_1, \cdots, a_Q\}$ of size Q. The QAM information symbols are arranged in the time-frequency plane (M subcarriers and N time slots), forming an information symbol matrix $\mathbf{X} \in \mathbb{C}^{M \times N}$, containing N OFDM symbols.

To simplify our explanation, we consider only *one OFDM symbol* (i.e., $N = 1$) in the rest of the chapter. Then, we denote a column of $\mathbf{X}$ as

$$\mathbf{x} = \left[\mathbf{x}[0], \ldots, \mathbf{x}[M-1]\right]^T \in \mathbb{C}^{M \times 1}$$

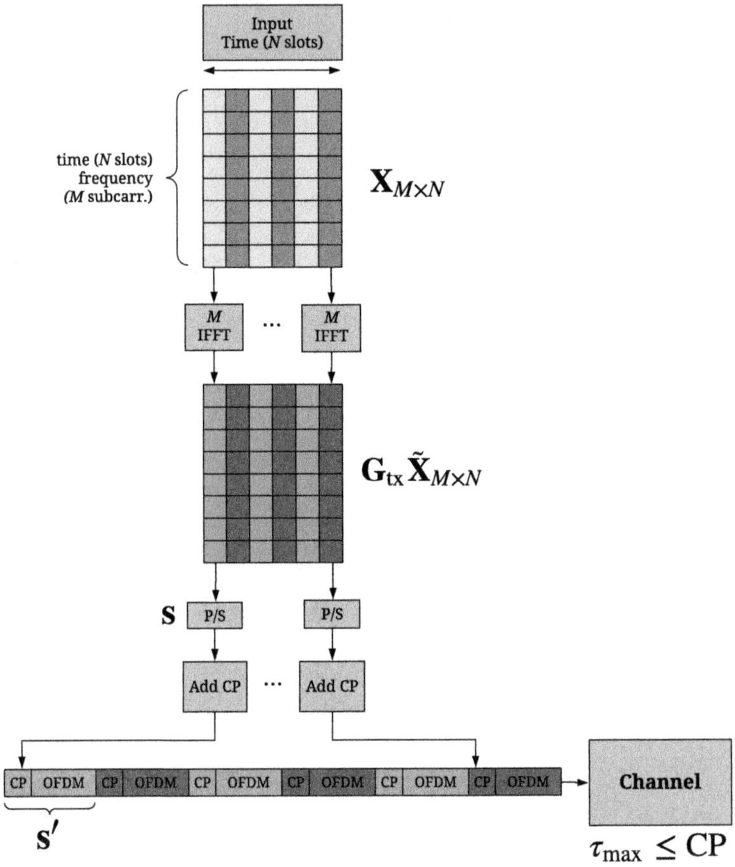

FIGURE 3.6 OFDM transmitter.

with element $\mathbf{x}[m] \in \mathbb{A}$, $m = 0, \ldots, M-1$.

An M-point inverse fast Fourier transform (IFFT), denoted by $\mathbf{F}_M^{\dagger}$, is applied to $\mathbf{x}$, yielding the time domain OFDM symbol vector as

$$\tilde{\mathbf{x}} = \left[\tilde{\mathbf{x}}[0], \ldots, \tilde{\mathbf{x}}[M-1]\right]^T = \mathbf{F}_M^{\dagger}\mathbf{x} \in \mathbb{C}^{M\times 1}. \tag{3.8}$$

Stacking these time domain OFDM symbol vectors forms the matrix $\tilde{\mathbf{X}} \in \mathbb{C}^{M\times N}$. Then, a pulse shaping waveform g_{tx} is applied to $\tilde{\mathbf{X}}$, yielding the time domain matrix $\mathbf{G}_{\text{tx}}\tilde{\mathbf{X}}$, where the diagonal matrix $\mathbf{G}_{\text{tx}}$ has the samples of $g_{\text{tx}}(t)$ as its entries as

$$\mathbf{G}_{\text{tx}} = \text{diag}\left[g_{\text{tx}}(0), g_{\text{tx}}(T/M), \ldots, g_{\text{tx}}\big((M-1)T/M\big)\right] \in \mathbb{C}^{M\times M}. \tag{3.9}$$

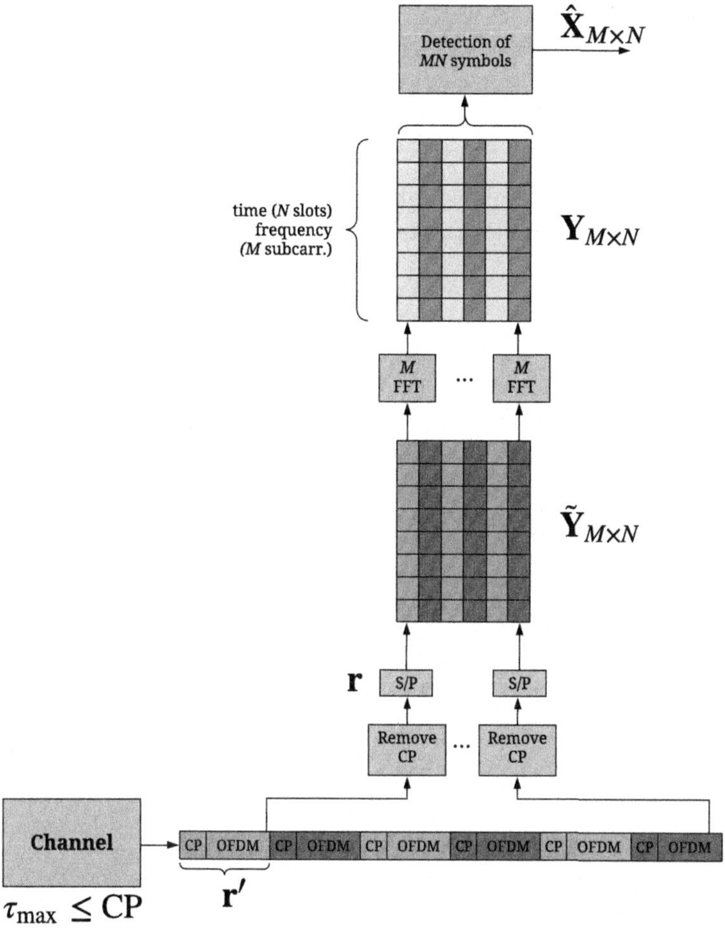

FIGURE 3.7 OFDM receiver.

For rectangular pulse shaping waveforms, it reduces to

$$\mathbf{G}_{\text{tx}} = \mathbf{I}_M \quad \text{and} \quad \mathbf{G}_{\text{tx}}\tilde{\mathbf{X}} = \tilde{\mathbf{X}}, \tag{3.10}$$

where $\mathbf{I}_M$ is the $M \times M$ identity matrix. After the parallel-to-serial conversion, each column of $\tilde{\mathbf{X}}$ is the OFDM symbol vector, given by

$$\mathbf{s} = \left[\mathbf{s}[0], \ldots, \mathbf{s}[M-1]\right]^T = \tilde{\mathbf{x}} \in \mathbb{C}^{M\times 1}. \tag{3.11}$$

A CP of length $L_{\text{CP}} = l_{\max}$, denoted by $\mathbf{s}_{\text{CP}} = [\mathbf{s}[M - l_{\max}], \ldots, \mathbf{s}[M-1]]^T$, is prepended to each $\mathbf{s}$ to obtain $\mathbf{s}' = [\mathbf{s}_{\text{CP}}^T, \ \mathbf{s}^T]$.

After digital-to-analog conversion and frequency up-conversion, the bandpass signal is transmitted over a static multipath channel.

3.3 OFDM frequency domain input–output relation

A static multipath channel with P distinct delay paths has a baseband equivalent channel impulse response, given by

$$h(t) = \sum_{i=1}^{P} g_i \delta(t - \tau_i), \tag{3.12}$$

where g_i and τ_i denote the complex path gain and the delay associated with the i-th path. When sampling at frequency f_s, we let $l_i = \lfloor \tau_i f_s \rfloor \in [0, l_{\max}]$ be the delay tap of the i-th path, for $i = 1, \ldots, P$. Then the discrete-time equivalent channel vector is given by $\mathbf{h}' = [h_0, \ldots, h_l, \ldots, h_{l_{\max}}]^T$, where

$$h_l = \begin{cases} g_i & \text{if } l = l_i, \\ 0 & \text{if } l \neq l_i, \end{cases} \quad i = 1, \ldots, P. \tag{3.13}$$

After down-conversion and analog-to-digital (AD) conversion (see Fig. 3.7), the equivalent baseband signal per slot (including CP) is given by

$$\mathbf{r}' = \mathbf{h}' * \mathbf{s}' + \mathbf{w}' \tag{3.14}$$

of length $M + l_{\max}$, where $\mathbf{w}'$ is the AWGN noise vector of the same length. After removing the first $l_{\max}$ samples from $\mathbf{r}'$ and serial-to-parallel (SP) conversion, we obtain the time domain received signal samples

$$\mathbf{r} = \left[\mathbf{r}[0], \ldots, \mathbf{r}[M-1]\right]^T = \left[\mathbf{r}'[l_{\max}], \ldots, \mathbf{r}'[M + l_{\max} - 1]\right]^T \in \mathbb{C}^{M \times 1}, \tag{3.15}$$

which is equivalent to the circular convoluation of the two vectors $\mathbf{h}$ and $\mathbf{s}$ of length M, i.e.,

$$\mathbf{r} = \mathbf{h} \circledast \mathbf{s} + \mathbf{w}, \tag{3.16}$$

where $\mathbf{h} = [\mathbf{h}[0], \ldots, \mathbf{h}[M-1]] = [h_0, \ldots, h_{l_{\max}}, 0, \ldots, 0]$ and $\mathbf{w}$ represents the AWGN noise vector of length M. Equivalently, in matrix form we have

$$\mathbf{r} = \mathbf{H}\mathbf{s} + \mathbf{w} \tag{3.17}$$

with

$$\mathbf{H} = \sum_{i=1}^{P} g_i \boldsymbol{\Pi}^{l_i} = \underbrace{\begin{bmatrix} h_0 & 0 & \cdots & 0 & h_{l_{\max}} & h_{l_{\max}-1} & \cdots & h_1 \\ h_1 & h_0 & \cdots & 0 & 0 & h_{l_{\max}} & \cdots & h_2 \\ \vdots & \ddots & \ddots & \ddots & \ddots & \ddots & \ddots & \vdots \\ h_{l_{\max}-1} & \ddots & \ddots & \ddots & \ddots & \ddots & 0 & h_{l_{\max}} \\ h_{l_{\max}} & h_{l_{\max}-1} & \ddots & \ddots & \ddots & \ddots & \ddots & 0 \\ 0 & h_{l_{\max}} & \ddots & \ddots & \ddots & \ddots & \ddots & \vdots \\ \vdots & \ddots & \ddots & \ddots & \ddots & \ddots & h_0 & 0 \\ 0 & 0 & \cdots & h_{l_{\max}} & h_{l_{\max}-1} & \cdots & h_1 & h_0 \end{bmatrix}}_{M \times M \text{ Circulant matrix}}, \tag{3.18}$$

where $\boldsymbol{\Pi}^{l_i}$ represents the l_i-th step circular shift of the permutation matrix $\boldsymbol{\Pi}$, as defined in Appendix B.

Given the eigenvalue/eigenvector decomposition of circulant matrices (see Appendix B.3), the channel matrix $\mathbf{H}$ is diagonalized as

$$\mathbf{H} = \sum_{i=1}^{P} g_i \boldsymbol{\Pi}^{l_i} = \mathbf{F}_M^{\dagger} \mathbf{D} \mathbf{F}_M, \tag{3.19}$$

and the diagonal matrix $\mathbf{D} \in \mathbb{C}^{M \times M}$ is given by

$$\mathbf{D} = \text{diag}(\mathbf{F}_M \mathbf{h}) = \text{diag}(\check{\mathbf{h}}), \tag{3.20}$$

where $\check{\mathbf{h}} = \mathbf{F}_M \mathbf{h} = [\check{\mathbf{h}}[0], \ldots, \check{\mathbf{h}}[M-1]]^T \in \mathbb{C}^{M \times 1}$ is the frequency domain channel response.

Substituting (3.19) into (3.17) yields the time domain input–output relation

$$\mathbf{r} = \mathbf{H}\mathbf{s} + \mathbf{w} = \mathbf{F}_M^{\dagger} \mathbf{D} \mathbf{F}_M \mathbf{s} + \mathbf{w}. \tag{3.21}$$

Collecting the column vector $\mathbf{r}$ from each time slot and applying the receiver rectangular pulse shaping waveform g_{rx} (its matrix form $\mathbf{G}_{\text{rx}} = \mathbf{I}_M$) yields the matrix $\tilde{\mathbf{Y}} \in \mathbb{C}^{M \times N}$. Applying the FFT operation on each column of $\tilde{\mathbf{Y}}$, we obtain the frequency domain received signal vector $\mathbf{y}$ (a column of matrix $\mathbf{Y} \in \mathbb{C}^{M \times N}$) as

$$\mathbf{y} = \mathbf{F}_M \mathbf{r} = \mathbf{D}\mathbf{x} + \check{\mathbf{w}} = \check{\mathbf{h}} \circ \mathbf{x} + \check{\mathbf{w}}, \tag{3.22}$$

where $\circ$ denotes element-wise multiplication, $\check{\mathbf{w}} = \mathbf{F}_M \mathbf{w}$, and $\mathbf{x} = \mathbf{F}_M \mathbf{s}$. Since $\mathbf{F}_M$ is a unitary matrix, the noise vector $\mathbf{w}$ remains independent and

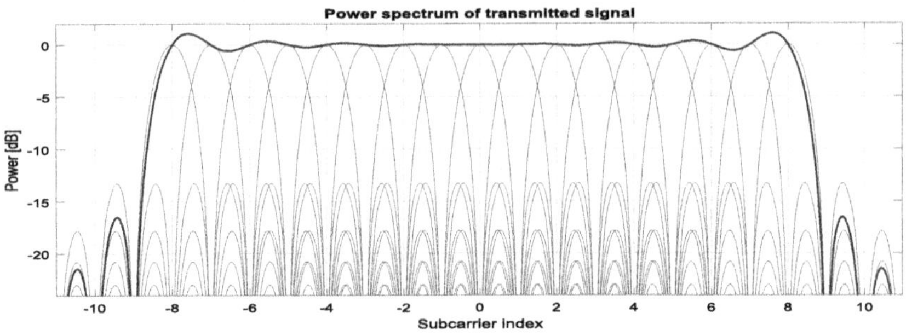

FIGURE 3.8 OFDM power spectrum.

identically distributed. Then a single tap equalizer can be adopted to estimate the QAM symbol vector (a column of matrix $\hat{\mathbf{X}} \in \mathbb{C}^{M \times N}$)

$$\hat{\mathbf{x}} = \mathcal{D}_{\mathbb{A}}\left(\mathbf{D}^{-1}\mathbf{y}\right), \tag{3.23}$$

where $\mathcal{D}_{\mathbb{A}}(\cdot)$ denotes the QAM symbol decision operation. Finally, a QAM demapper can be used to recover the information bits.

Assuming the channel is static over multiple OFDM symbols, the channel coefficients $\check{\mathbf{h}}[m]$, $m = 0, \ldots, M - 1$, can be easily estimated using $\mathbf{y}[m]$ by transmitting pilot symbol $\mathbf{x} = \mathbf{1}_{M \times 1}$.

3.4 Advantages and disadvantages of OFDM

In summary, OFDM has the following key features:

- OFDM adopts multiple subcarriers to carry information, where all subcarriers are orthogonal to each other under certain conditions;
- a CP of length $l_{\max}$ is prepended to each OFDM symbol to combat severe multipath channel delay spread and ISI; and
- the CP provides a circulant channel matrix structure as in (3.18), which enables single tap equalization to combat severe multipath channel impairments.

Overall, OFDM enables a low complexity modulation and demodulation thanks to the FFT and the CP. On the other hand, OFDM has some disadvantages that need to be considered.

3.4.1 High PAPR

The PAPR is the ratio between the maximum power of a time domain sample in an OFDM transmit symbol and the average power of that

OFDM symbol, defined as

$$\text{PAPR}_{\text{dB}} \triangleq 10\log_{10}\frac{\max\{|\mathbf{s}[m]|^2\}}{\text{E}\{|\mathbf{s}[m]|^2\}}, \quad m=0,\ldots,M-1, \tag{3.24}$$

where

$$\mathbf{s}[m]=\frac{1}{\sqrt{M}}\sum_{k=0}^{M-1}\mathbf{x}[k]e^{\frac{j2\pi km}{M}}.$$

Note that $\text{E}(|\mathbf{x}[k]|^2)=E_s$, $\text{E}(\mathbf{x}[k])=0$, for all k, and

$$\begin{aligned}\text{E}\big(|\mathbf{s}[m]|^2\big)&=\text{E}\left(\frac{1}{M}\sum_{k=0}^{M-1}\sum_{k'=0}^{M-1}\mathbf{x}[k]\mathbf{x}^*\big[k'\big]e^{\frac{j2\pi(k-k')m}{M}}\right)\\&=\frac{1}{M}\sum_{k=0}^{M-1}\sum_{k'=0}^{M-1}\text{E}\big(\mathbf{x}[k]\mathbf{x}^*\big[k'\big]\big)e^{\frac{j2\pi(k-k')m}{M}}\\&=\frac{1}{M}\left(\sum_{k=0}^{M-1}\underbrace{\text{E}\big(|\mathbf{x}[k]|^2\big)}_{=E_S}+\sum_{k=0}^{M-1}\sum_{k'=0,k'\neq k}^{M-1}\underbrace{\text{E}\big(\mathbf{x}[k]\big)\text{E}\big(\mathbf{x}^*\big[k'\big]\big)}_{=0}e^{\frac{j2\pi(k-k')m}{M}}\right)\\&=E_s.\end{aligned} \tag{3.25}$$

Assuming $|\mathbf{x}[k]|^2=A^2$, for all k, where A^2 denotes the *peak power constellation point* with

$$A^2=\alpha E_s \text{ and } \alpha\geq 1,$$

we obtain

$$\max\big(|\mathbf{s}[m]|^2\big)=|\mathbf{s}[0]|^2=\left|\frac{1}{\sqrt{M}}\sum_{k=0}^{M-1}\mathbf{x}[k]\right|^2=MA^2. \tag{3.26}$$

Hence, (3.24) becomes

$$\text{PAPR}_{\text{dB}}=\frac{MA^2}{E_s}=\alpha M. \tag{3.27}$$

Note that as $\alpha\geq 1$ when M is a large number, the PAPR value can be very high. This is incurred by the IFFT operation, i.e., data symbols across subcarriers can add up to produce a peak value signal. High peaks can cause the power amplifier to operate in the nonlinear region, leading to degraded system performance. In summary, OFDM PAPR reduction is a critical issue and many PAPR reduction techniques have been developed in the past two decades, for example, clipping and filtering, single carrier frequency division multiple access (SC-FDMA), phase optimization, tone reservation, and constellation shaping.

3.4.2 High OOB

At a fixed time slot, given the normalized angular frequency $\hat{\omega}$ and assuming no CP, the total instantaneous spectral density of the signal at the output of the OFDM system is given by

$$S(\hat{\omega}) = \sum_{m=0}^{M-1} H_m(\hat{\omega}) S_{\mathbf{x}[m]}(\hat{\omega}), \tag{3.28}$$

where $S_{\mathbf{x}[m]}(\hat{\omega})$ is the corresponding spectral density of the QAM symbol $\mathbf{x}[m]$ and $H_m(\omega_n)$ is the m-th subcarrier shaping filter frequency response

$$H_m(\hat{\omega}) = \frac{\sin(M(\frac{\hat{\omega}}{2} - \frac{\pi m}{M}))}{M \sin(\frac{\hat{\omega}}{2} - \frac{\pi m}{M})}. \tag{3.29}$$

Assuming uncorrelated QAM symbols with unit energy, Fig. 3.8 illustrates the power spectral density (PSD) $|S(\hat{\omega})|^2$ of OFDM signals without CP and demonstrates a high OOB radiation (-17 dB sidelobes), which can disrupt communications in adjacent wireless channels. The OOB spectrum of OFDM decreases slowly according to a sinc function with second lobe attenuation of 13 dB for one OFDM symbol with $M = 20$. A large number of OOB suppression techniques have been studied in the literature, such as filtering and windowing, carrier cancelations, subcarrier weighting, multiple choice sequence technique, and various precoding techniques.

3.4.3 Sensitivity to CFO

OFDM is sensitive to frequency offset, caused by a deviation between the transmitter and receiver frequencies, or due to Doppler shift as the transmitter or the receiver is moving. In such cases, the received signal is shifted in frequency and the sampling in the frequency domain does not coincide with the center frequencies of the subcarriers. Therefore the amplitude of the desired subcarriers will decrease and intercarrier interference (ICI) will appear.

Let us define the frequency offset as

$$f_o = f_c - f_c', \tag{3.30}$$

where f_c and f_c' are the carrier frequencies in the transmitter and the receiver, respectively. The normalized carrier frequency offset is defined as

$$\epsilon = \frac{f_o}{\Delta f} = \lfloor \epsilon \rfloor + \Delta\epsilon, \tag{3.31}$$

where $\lfloor \epsilon \rfloor$ and $\Delta\epsilon$ are the integer and fractional parts of ϵ, respectively.

Considering a noiseless case for simplicity, (3.22) becomes

$$\mathbf{y}^{\text{cfo}} = \mathbf{C}\mathbf{D}\mathbf{x}, \tag{3.32}$$

where $\mathbf{C} = \{c_{m,m'}\}_{m,m'=0}^{M-1}$ is a Toeplitz matrix that represents the frequency domain convolution operation due to CFO and

$$c_{m,m'} = \frac{\sin(\pi(m-m'+\epsilon))}{M\sin(\frac{\pi(m-m'+\epsilon)}{M})} e^{\frac{j\pi(m-m'+\epsilon)(M-1)}{M}} \qquad m, m' \in [0, M-1]. \tag{3.33}$$

Note that the CFO matrix $\mathbf{C}$ has diagonal elements

$$c_{m,m} = \frac{\sin(\pi\epsilon)}{M\sin(\frac{\pi\epsilon}{M})} e^{\frac{j\pi\epsilon(M-1)}{M}} \qquad m \in [0, M-1]. \tag{3.34}$$

From (3.34), we can see that the CFO incurs degradation of the amplitude of the signal at subcarrier $m = m'$ by a factor of

$$\frac{\sin(\pi\epsilon)}{M\sin(\frac{\pi\epsilon}{M})}$$

and a phase shift due to the term $e^{\frac{j\pi\epsilon(M-1)}{M}}$. Moreover, the signal suffers from ICI due to $c_{m,m'} \neq 0$ for $m \neq m'$ as shown in (3.33). A receiver will need to estimate the CFO $f_o = f_c - f_c'$ using a pilot tone and then compensate for it by multiplying the output of the mixer by $e^{-j2\pi \hat{f}_o t}$.

3.5 OFDM in high-mobility multipath channels

In this section, we consider OFDM in a high-mobility multipath channel, where each path has different Doppler shifts $\nu_i = \kappa_i / T$ for $i = 1, \ldots, P$ and $k_i = \lfloor \kappa_i \rceil$ is the Doppler tap. The time domain received signal samples are given by

$$\mathbf{r} = \left[\mathbf{r}(0), \ldots, \mathbf{r}(M-1)\right]^T = \mathbf{H}\mathbf{s} + \mathbf{w} \in \mathbb{C}^{M\times 1}, \tag{3.35}$$

where

$$\mathbf{H} = \sum_{i=1}^{P} g_i \underbrace{\boldsymbol{\Pi}^{l_i}}_{\text{delay}} \underbrace{\boldsymbol{\Delta}^{k_i}}_{\text{Doppler}} \tag{3.36}$$

$$
= \begin{bmatrix}
h_0 & 0 & \cdots & h_{l_{\max}}\omega^{k_P(M-l_{\max}))} & \cdots & h_1\omega^{k_1(M-1)} \\
\vdots & h_0\omega^{k_0} & \cdots & \ddots & \cdots & \vdots \\
\vdots & \ddots & \ddots & 0 & \ddots & \vdots \\
\vdots & \ddots & \ddots & \ddots & & h_{l_{\max}}\omega^{k_P(M-1)} \\
h_{l_{\max}} & \ddots & \ddots & \ddots & \ddots & 0 \\
0 & h_{l_{\max}}\omega^{k_P} & \ddots & \ddots & \ddots & \vdots \\
\vdots & \ddots & \ddots & \ddots & \ddots & 0 \\
0 & \cdots & h_{l_{\max}}\omega^{k_P(M-l_{\max}-1)} & \cdots & \cdots & h_0\omega^{k_0(M-1)}
\end{bmatrix},
$$

where $\boldsymbol{\Pi}^{l_i}$ represents the l_i-th step circular shift of the permutation matrix $\boldsymbol{\Pi}$, as defined in Appendix B, and $\boldsymbol{\Delta}^{k_i}$ is an $M \times M$ diagonal matrix, given by

$$\boldsymbol{\Delta}^{k_i} = \text{diag}\big(1, \omega^{k_i}, \cdots, \omega^{k_i(M-1)}\big). \tag{3.37}$$

The matrices $\boldsymbol{\Pi}^{l_i}$ and $\boldsymbol{\Delta}^{k_i}$ model the delays and the Doppler shifts in the l_i-th delay path. Multiplying the channel matrix $\mathbf{H}$ by $\mathbf{s}$ implies that the l_i-th path introduces an l_i-step cyclic shift of the transmitted signal vector $\mathbf{s}$, modeled by $\boldsymbol{\Pi}^{l_i}$, and modulates it with a carrier at frequency k_i, modeled by $\boldsymbol{\Delta}^{k_i}$. Multiple Dopplers introduce $\boldsymbol{\Delta}^{k_i}$ to $\mathbf{H}$, making $\mathbf{H}$ *no longer a circulant matrix*. As a result, the eigenvalue decomposition of $\mathbf{H}$ (see (3.19)) does not hold and severe ICI appears.

In summary, multiple Dopplers challenge OFDM, since multiple Dopplers are hard to equalize, and subchannel gains are not equal and the lowest gain decides the performance (see (3.36)). Hence, new modulation techniques are required to combat high Dopplers.

3.6 Bibliographical notes

Readers can refer to [1–14] for the basic concepts of OFDM and the analysis of PAPR reduction techniques in OFDM. Details regarding OOB reduction methods including windowing, filtering, and precoding can be found in [13,15–32].

References

[1] X. Li, L. Cimini, Effects of clipping and filtering on the performance of OFDM, IEEE Communications Letters 2 (5) (1998) 131–133, https://doi.org/10.1109/4234.673657.

[2] R. O'Neill, L. Lopes, Envelope variations and spectral splatter in clipped multicarrier signals, in: Proceedings of 6th International Symposium on Personal, Indoor and Mobile Radio Communications, vol. 1, 1995, pp. 71–75, https://doi.org/10.1109/PIMRC.1995.476406.

[3] H. Ochiai, H. Imai, On clipping for peak power reduction of OFDM signals, in: Globecom'00 – IEEE. Global Telecommunications Conference. Conference Record (Cat. No. 00CH37137), vol. 2, 2000, pp. 731–735, https://doi.org/10.1109/GLOCOM.2000.891236.

[4] F. Nadal, S. Sezginer, H. Sari, Peak-to-average power ratio reduction in CDMA systems using metric-based symbol predistortion, IEEE Communications Letters 10 (8) (2006) 577–579, https://doi.org/10.1109/LCOMM.2006.1665115.

[5] H.G. Myung, J. Lim, D.J. Goodman, Peak-to-average power ratio of single carrier FDMA signals with pulse shaping, in: 2006 IEEE 17th International Symposium on Personal, Indoor and Mobile Radio Communications, 2006, pp. 1–5.

[6] G.L. Li, Y.G. Stuber, Orthogonal Frequency Division Multiplexing for Wireless Communications, 1st edition, Springer US, USA, 2006.

[7] A. Jones, T. Wilkinson, S.K. Barton, Block coding scheme for reduction of peak to mean envelope power ratio of multicarrier transmission scheme, Electronics Letters 30 (22) (1994) 2098–2099.

[8] A. Jones, T. Wilkinson, Combined coding for error control and increased robustness to system nonlinearities in OFDM, in: Proceedings of Vehicular Technology Conference – VTC, vol. 2, 1996, pp. 904–908, https://doi.org/10.1109/VETEC.1996.501442.

[9] Z.G. Tao, J. Zheng, Block coding scheme for reducing PAPR in OFDM systems with large number of subcarriers, Journal of Electronics 21 (6) (2004) 482–489.

[10] J. Tellado, Multicarrier Modulation with Low Peak to Average Power: Applications to xDSL and Broadband Wireless, Springer US, 2000.

[11] B. Krongold, D. Jones, An active-set approach for OFDM PAR reduction via tone reservation, IEEE Transactions on Signal Processing 52 (2) (2004) 495–509, https://doi.org/10.1109/TSP.2003.821110.

[12] H. Kwok, D. Jones, PAPR reduction via constellation shaping, in: 2000 IEEE International Symposium on Information Theory (Cat. No.00CH37060), 2000, p. 166, https://doi.org/10.1109/ISIT.2000.866461.

[13] H.A. Mahmoud, H. Arslan, Sidelobe suppression in OFDM-based spectrum sharing systems using adaptive symbol transition, IEEE Communications Letters 12 (2) (2008) 133–135, https://doi.org/10.1109/LCOMM.2008.071729.

[14] D. Jones, Peak power reduction in OFDM and DMT via active channel modification, in: Conference Record of the Thirty-Third Asilomar Conference on Signals, Systems, and Computers (Cat. No.CH37020), vol. 2, 1999, pp. 1076–1079, https://doi.org/10.1109/ACSSC.1999.831875.

[15] M. Faulkner, The effect of filtering on the performance of OFDM systems, IEEE Transactions on Vehicular Technology 49 (5) (2000) 1877–1884, https://doi.org/10.1109/25.892590.

[16] Y.-P. Lin, S.-M. Phoong, Window designs for DFT-based multicarrier systems, IEEE Transactions on Signal Processing 53 (3) (2005) 1015–1024, https://doi.org/10.1109/TSP.2004.842173.

[17] J. Abdoli, M. Jia, J. Ma, Filtered OFDM: a new waveform for future wireless systems, in: 2015 IEEE 16th International Workshop on Signal Processing Advances in Wireless Communications (SPAWC), 2015, pp. 66–70.

[18] S. Brandes, I. Cosovic, M. Schnell, Reduction of out-of-band radiation in OFDM systems by insertion of cancellation carriers, IEEE Communications Letters 10 (6) (2006) 420–422, https://doi.org/10.1109/LCOMM.2006.1638602.

[19] J.F. Schmidt, S. Costas-Sanz, R. López-Valcarce, Choose your subcarriers wisely: active interference cancellation for cognitive OFDM, IEEE Journal on Emerging and Selected Topics in Circuits and Systems 3 (4) (2013) 615–625, https://doi.org/10.1109/JETCAS.2013.2280808.

[20] I. Cosovic, S. Brandes, M. Schnell, Subcarrier weighting: a method for sidelobe suppression in OFDM systems, IEEE Communications Letters 10 (6) (2006) 444–446, https://doi.org/10.1109/LCOMM.2006.1638610.

[21] I. Cosovic, T. Mazzoni, Suppression of sidelobes in OFDM systems by multiple-choice sequences, European Transactions on Telecommunications 17 (6) (2006) 623–630.
[22] D. Li, X. Dai, H. Zhang, Sidelobe suppression in NC-OFDM systems using constellation adjustment, IEEE Communications Letters 13 (5) (2009) 327–329, https://doi.org/10.1109/LCOMM.2009.090031.
[23] R. Xu, M. Chen, J. Zhang, B. Wu, H. Wang, Spectrum sidelobe suppression for discrete Fourier transformation-based orthogonal frequency division multiplexing using adjacent subcarriers correlative coding, IET Communications 6 (11) (2012) 1374–1381, https://doi.org/10.1049/iet-com.2011.0007.
[24] X. Huang, J.A. Zhang, Y.J. Guo, Out-of-band emission reduction and a unified framework for precoded OFDM, IEEE Communications Magazine 53 (6) (2015) 151–159, https://doi.org/10.1109/MCOM.2015.7120032.
[25] M. Ma, X. Huang, B. Jiao, Y.J. Guo, Optimal orthogonal precoding for power leakage suppression in DFT-based systems, IEEE Transactions on Communications 59 (3) (2011) 844–853, https://doi.org/10.1109/TCOMM.2011.121410.100071.
[26] A. Tom, A. Sahin, H. Arslan, Mask compliant precoder for OFDM spectrum shaping, IEEE Communications Letters 17 (3) (2013) 447–450, https://doi.org/10.1109/LCOMM.2013.020513.122495.
[27] J. van de Beek, F. Berggren, N-continuous OFDM, IEEE Communications Letters 13 (1) (2009) 1–3, https://doi.org/10.1109/LCOMM.2009.081446.
[28] J. Van De Beek, Sculpting the multicarrier spectrum: a novel projection precoder, IEEE Communications Letters 13 (12) (2009) 881–883, https://doi.org/10.1109/LCOMM.2009.12.091614.
[29] J.A. Zhang, X. Huang, A. Cantoni, Y.J. Guo, Sidelobe suppression with orthogonal projection for multicarrier systems, IEEE Transactions on Communications 60 (2) (2012) 589–599, https://doi.org/10.1109/TCOMM.2012.012012.110115.
[30] Y. Zheng, J. Zhong, M. Zhao, Y. Cai, A precoding scheme for N-continuous OFDM, IEEE Communications Letters 16 (12) (2012) 1937–1940, https://doi.org/10.1109/LCOMM.2012.102612.122168.
[31] X. Zhou, G.Y. Li, G. Sun, Multiuser spectral precoding for OFDM-based cognitive radio systems, IEEE Journal on Selected Areas in Communications 31 (3) (2013) 345–352, https://doi.org/10.1109/JSAC.2013.130302.
[32] L. Pan, J. Ye, X. Yuan, Spectral precoding for out-of-band power reduction under condition number constraint in OFDM-based systems, Wireless Personal Communications 95 (2) (2017) 1677–1691, https://doi.org/10.1007/s11277-016-3874-8.

CHAPTER

4

Delay-Doppler modulation

OUTLINE

Delay-Doppler Communications

https://doi.org/10.1016/B978-0-32-385028-5.00012-8

Chapter points

- OTFS modulation and demodulation.
- OTFS matrix formulation.
- The discrete Zak transform.
- OTFS input–output relations in different domains.
- Variants of OTFS.

Non c'è nulla di più difficile da gestire, di esito incerto e così pericoloso da realizzare dell'inizio di un cambiamento. **Niccolò Machiavelli**

With the advent of high speed trains, unmanned aerial vehicles (UAVs), and self-driving cars, there is an urgent need for reliable communications in high-mobility wireless channels. In OFDM, information symbols are transmitted over a single time-frequency resource, which is susceptible to frequency and time selective fading effects that degrade the error performance in high-mobility wireless channels. On the other hand, OTFS multiplexes each information symbol over a two-dimensional (2D) orthogonal basis function that spans the entire time and frequency resources. As a result, all information symbols experience a fixed (time-invariant) flat fading equivalent channel.

This chapter starts with some basic notations about different domains involved, such as the discrete-time domain, time-frequency domain, and delay-Doppler domain, followed by the description of OTFS modulation and demodulation, high-mobility channels, and OTFS input–output relations with ideal pulse shaping waveforms.

Later in the chapter, we introduce OTFS matrix forms, demonstrating how the OTFS modulation is related to the well-known discrete Zak transform. Then, we present input–output relations in vectorized form for OTFS with practical rectangular pulse shaping waveforms, where different domains are considered: the discrete-time domain, time-frequency domain, delay-time domain, and delay-Doppler domain. Further, we extend our study to the variants of OTFS, where the waveform is transmitted with a cyclic prefix (CP) or zero padding (ZP) added to each OTFS frame or block. Finally, we present a comprehensive summary of channel representations and input–output relations for the variants of OTFS.

4.1 System model

In this section, prior to describing the OTFS modulation and demodulation blocks in Fig. 4.1, we introduce some notations about the three domains involved: the discrete-time domain, time-frequency domain, and delay-Doppler domain.

We assume that the OTFS system operates on a P-path high-mobility channel with a bandwidth B, maximum delay spread τ_{max}, and maximum Doppler shift ν_{max}, defined in (2.11). We consider a discrete-time baseband equivalent model, where a continuous-time OTFS signal is sampled at a sampling frequency $f_s = B = \frac{1}{T_s}$, where T_s denotes the sampling interval. The discrete-time domain OTFS frame contains NM samples subdivided into N blocks (or time slots), with M samples per block. Hence, the OTFS frame duration is $T_{\text{f}} = NMT_s = NT$, where $T = MT_s$ denotes the duration of each block.

Every T seconds, we obtain the discrete spectrum of each block resulting from an M-point discrete Fourier transform (DFT), where the spectrum samples are spaced by $\Delta f = 1/T$. Collecting all N spectra of bandwidth $B = M\Delta f$ of the OTFS frame along the time axis defines the discrete *time-frequency domain*, as shown in Fig. 4.2(left). The discrete time-frequency domain is defined as the $M \times N$ array of points

$$\Lambda = \left\{(l\Delta f, kT),\ l = 0, \ldots, M-1, k = 0, \ldots, N-1\right\}$$

for integers $M, N > 0$. The discrete time-frequency samples at the points in Λ are collected in the matrix $\mathbf{X}_{\text{tf}}[l, k]$, $l = 0, \ldots, M-1$, $k = 0, \ldots, N-1$, where each column contains the discrete spectrum samples of each block. It is convenient to think of this matrix as a 2D time-frequency representation of the 1D time domain OTFS signal.

Discrete time-frequency samples can be converted to the *delay-Doppler domain* via a 2D symplectic Fourier transform. Specifically, the delay-Doppler domain is obtained from the time-frequency domain by an inverse Fourier transform along the frequency axis (columns of $\mathbf{X}_{\text{tf}}[l,k]$) and a Fourier transform along the time axis (rows of $\mathbf{X}_{\text{tf}}[l,k]$). When discretized, the corresponding $M \times N$ array of points in the delay-Doppler domain (see Fig. 4.2(right)) is

$$\Gamma = \left\{ \left(\frac{m}{M\Delta f}, \frac{n}{NT} \right), \; m = 0, \ldots, M-1, n = 0, \ldots, N-1 \right\}, \tag{4.1}$$

where $\frac{1}{M\Delta f}$ and $\frac{1}{NT}$ are the resolutions of the path delays and the Doppler shifts, respectively. In particular, two paths with the same Doppler shift but with a difference in propagation delay less than $\frac{1}{M\Delta f}$ cannot be distinguished by the receiver. Similarly, two paths with the same propagation delay but with a difference in Doppler shift less than $\frac{1}{NT}$ also cannot be distinguished.

We define the delay-Doppler samples of the OTFS waveform at the points in Γ as the matrix $\mathbf{X}[m,n], m = 0, \ldots, M-1, n = 0, \ldots, N-1$.

4.1.1 Parameter choice for OTFS systems

As key design parameters, we may choose N, M, and T (since $\Delta f = 1/T$). We can see that T and Δf determine the maximum supportable channel delay $\tau_{\max} < T$ and Doppler shifts $\nu_{\max} < \Delta f/2$. If we fix the data rate to NM symbols per frame, depending on the channel parameters, we can choose a larger T (smaller Δf), resulting in a smaller N and larger M, or vice versa. This means that OTFS can handle channels up to $\tau_{\max}\nu_{\max} < 1/2$, which expands the opportunities to design systems working beyond underspread channels ($\tau_{\max}\nu_{\max} \ll 1$). Another design constraint for OTFS systems lies in the assumption that the multipath channel parameters g_i, τ_i, and ν_i (see Chapter 2) are constant over the duration of the frame. This limits T_{f} to a maximum of 10–20 ms in today's cellular system environment.

4.1.2 OTFS modulation

As illustrated in Fig. 4.1, at the transmitter, the NM information symbols, taken from a modulation alphabet $\mathbb{A} = \{a_1, \cdots, a_Q\}$ of size Q, are placed in the *delay-Doppler domain matrix* $\mathbf{X} \in \mathbb{C}^{M \times N}$ with entries $\mathbf{X}[m,n]$, for $m = 0, \ldots, M-1, n = 0, \ldots, N-1$.

The transmitter first maps symbols $\mathbf{X}[m,n]$ to NM samples $\mathbf{X}_{\text{tf}}[l,k]$ on the time-frequency grid Λ via inverse symplectic fast Fourier transform

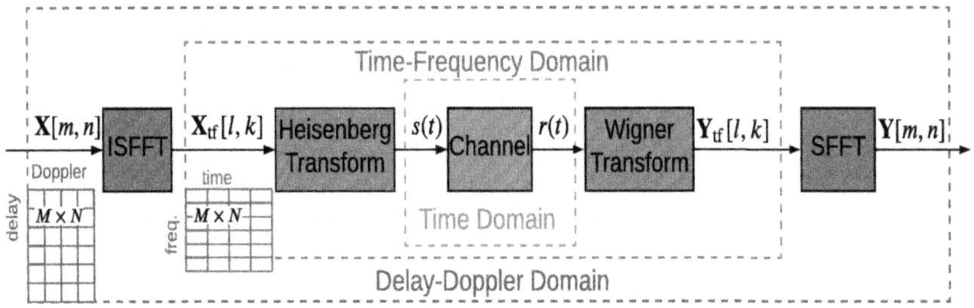

FIGURE 4.1 OTFS system diagram in its original form.

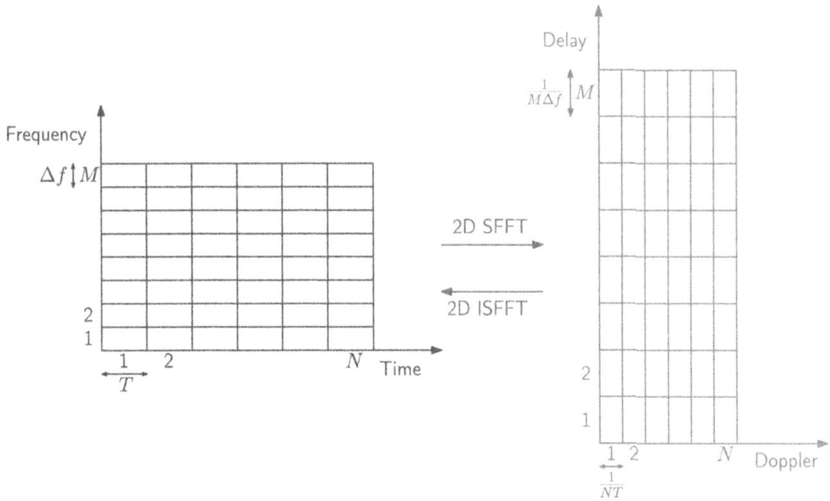

FIGURE 4.2 The discrete time-frequency grid (Λ) and delay-Doppler grid (Γ).

(ISFFT), i.e.,

$$\mathbf{X}_{\text{tf}}[l,k] = \frac{1}{\sqrt{NM}} \sum_{n=0}^{N-1} \sum_{m=0}^{M-1} \mathbf{X}[m,n] e^{j2\pi(\frac{nk}{N} - \frac{ml}{M})} \tag{4.2}$$

for $l = 0, \ldots, M-1$, $k = 0, \ldots, N-1$, where $\mathbf{X}_{\text{tf}} \in \mathbb{C}^{M \times N}$ represents the *time-frequnecy domain transmitted samples matrix*. The ISFFT corresponds to a 2D transformation which takes an M-point DFT of the columns of $\mathbf{X}$ and an N-point inverse DFT (IDFT) of the rows of $\mathbf{X}$.

Next, a time-frequency modulator converts the 2D samples $\mathbf{X}_{\text{tf}}[l,k]$ to a continuous-time waveform $s(t)$ using a transmit waveform $g_{\text{tx}}(t)$ as

$$\boxed{s(t)=\sum_{k=0}^{N-1}\sum_{l=0}^{M-1}\mathbf{X}_{\text{tf}}[l,k]g_{\text{tx}}(t-kT)e^{j2\pi l\Delta f(t-kT)}.} \tag{4.3}$$

The above operation is referred to in the literature as the Heisenberg transform, which depends on N, M, and $g_{\text{tx}}(t)$.

4.1.3 High-mobility channel distortion

The signal $s(t)$ is transmitted over a time-varying channel with delay-Doppler channel response $h(\tau,\nu)$, corresponding to the delay-time response $g(\tau,t)$, where τ,ν are the channel delay and Doppler shift (see Chapter 2). Omitting the noise term, the received signal $r(t)$ is given by

$$\begin{aligned} r(t) &= \iint h(\tau,\nu)s(t-\tau)e^{j2\pi\nu(t-\tau)}d\tau d\nu \\ &= \int g(\tau,t)s(t-\tau)d\tau, \end{aligned} \tag{4.4}$$

where $g(\tau,t)=\int_{\nu}h(\tau,\nu)e^{j2\pi\nu(t-\tau)}d\nu$ is given in (2.15).

At the receiver, the channel impaired signal is discretized by sampling $r(t)$ at $t=qT_s=qT/M$ for $q=0,\ldots,NM-1$, and $\tau=lT_s=lT/M$ for $l=0,\ldots,M-1$. Then (4.4) becomes

$$r[q]=\sum_{l}g^{\text{s}}[l,q]s[q-l],\quad q=0,\ldots,NM-1, \tag{4.5}$$

where the discrete delay-time response $g^{\text{s}}[l,q]$ was given in (2.19) as

$$g^{\text{s}}[l,q]=\sum_{\ell\in\mathcal{L}}\left(\sum_{\kappa\in\mathcal{K}_\ell}\nu_\ell(\kappa)z^{\kappa(q-l)}\right)\text{sinc}(l-\ell), \tag{4.6}$$

where $\text{sinc}(x)=\sin(\pi x)/(\pi x)$, $z=e^{\frac{j2\pi}{NM}}$, and $\nu_\ell(\kappa)$ is the Doppler response at delay shift $\ell T/M$. The set $\mathcal{L}$ contains the distinct normalized delay shifts ℓ of the channel and $\mathcal{K}_\ell$ is the set of normalized Doppler shifts κ of all the paths sharing the same delay shift $\ell T/M$.

Now we consider a time-varying multipath channel of P paths, where the i-th path ($i=1,\ldots,P$) has channel gain g_i, delay shift τ_i, and Doppler shift ν_i, the Doppler response in (2.13) at delay shift $\ell=\ell_i=\tau_i/M\Delta f$ is given as

$$\nu_\ell(\kappa)=\begin{cases} g_i, & \text{if } \ell=\ell_i=\tau_i/M\Delta f \text{ and } \kappa=\kappa_i=\nu_i/NT, \\ 0, & \text{otherwise.}\end{cases} \tag{4.7}$$

In paritcular, when the normalized delay shifts ℓ_i and the normalized Doppler shifts κ_i are integers, we denote them with l_i (*integer delay taps*), and k_i (*integer Doppler taps*). Then the sinc function in (4.6) is replaced by the unit pulse at l_i yielding the discrete delay-time multipath channel response

$$g^{s}[l,q]=\sum_{i=1}^{P}\nu_{l_i}(\kappa_i)z^{\kappa_i(q-l)}\delta[l-l_i]=\sum_{i=1}^{P}g_i z^{\kappa_i(q-l)}\delta[l-l_i] \qquad (4.8)$$

where the second step is due to (4.7). The maximum value of l_i is assumed to be less than the *maximum channel delay tap* $l_{\max}$, where $l_{\max} < M$.

It is important to remark here that the delay-Doppler (or equivalently the delay-time) response is assumed to be fixed over the duration of one OTFS frame. This means that the delay-Doppler channel parameters g_i, τ_i, and ν_i are assumed to be constant for $T_f = NT$. To satisfy this assumption, it may be necessary to reduce N at the cost of a lower Doppler shift resolution.

4.1.4 OTFS demodulation

In Fig. 4.1, at the receiver, the received signal $r(t)$ is passed through a matched filter computing the cross-ambiguity function $A_{g_{\mathrm{rx}},r}(f,t)$ as

$$Y(f,t)=A_{g_{\mathrm{rx}},r}(f,t)\triangleq\int r(t')g^{*}_{\mathrm{rx}}(t'-t)e^{-j2\pi f(t'-t)}dt' \qquad (4.9)$$

and then sampling $Y(f,t)$ on the grid points Λ forms the *time-frequency domain received samples matrix* $\mathbf{Y}_{\mathrm{tf}} \in \mathbb{C}^{M\times N}$ with entries

$$\mathbf{Y}_{\mathrm{tf}}[l,k]=Y(f,t)|_{f=l\Delta f,\ t=kT} \qquad (4.10)$$

for $l=0,\ldots,M-1$ and $k=0,\ldots,N-1$. Jointly, (4.9) and (4.10) are referred to as the *Wigner transform*.

Finally, the sympletic fast Fourier transform (SFFT) is applied on $\mathbf{Y}_{\mathrm{tf}}[m,n]$ to obtain the delay-Doppler domain samples as

$$\boxed{\mathbf{Y}[m,n]=\frac{1}{\sqrt{NM}}\sum_{k=0}^{N-1}\sum_{l=0}^{M-1}\mathbf{Y}_{\mathrm{tf}}[l,k]e^{-j2\pi\left(\frac{nk}{N}-\frac{ml}{M}\right)},} \qquad (4.11)$$

which form the *delay-Doppler domain received samples matrix* $\mathbf{Y} \in \mathbb{C}^{M\times N}$. The SFFT corresponds to a 2D transformation which takes an M-point IDFT of the columns of $\mathbf{Y}$ and an N-point DFT of the rows of $\mathbf{Y}$.

In summary, as shown in Fig. 4.1, the OTFS modulator maps $\mathbf{X}[m,n]$ in the delay-Doppler domain to $\mathbf{X}_{\mathrm{tf}}[l,k]$ in the time-frequency domain using

ISFFT. Then the Heisenberg transform is applied to $\mathbf{X}_{\text{tf}}[l,k]$ to generate time domain signal $s(t)$. At the receiver, $r(t)$ is transformed to the time-frequency domain through the Wigner transform and then to the delay-Doppler domain using SFFT, prior to symbol demodulation.

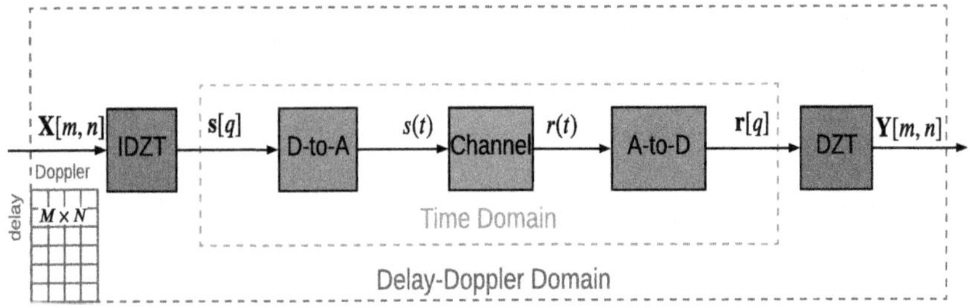

FIGURE 4.3 OTFS system diagram using the discrete Zak transform.

Remark. Alternatively, the OTFS transmitter can be realized using an inverse discrete Zak transform (IDZT) and a digital-to-analog (DA) converter to form the transmitted signal $s(t)$, as shown in Fig. 4.3. The OTFS receiver can be realized using an analog-to-digital (AD) converter in the received signal $r(t)$ followed by a discrete Zak transform (DZT). Such equivalence will become apparent in Section 4.3 and we refer the reader to Chapter 5 for details about the Zak transform.

4.2 OTFS input–output relation with ideal waveforms

The $g_{\text{rx}}(t)$ and $g_{\text{tx}}(t)$ pulses are said to be ideal if they satisfy the *biorthogonal property*:

$$A_{g_{\text{rx}},g_{\text{tx}}}(f,t)|_{f=m\Delta f+(-\nu_{\max},\nu_{\max}),t=nT+(-\tau_{\max},\tau_{\max})} = \delta[m]\delta[n]p_{\nu_{\max}}(f)p_{\tau_{\max}}(t), \tag{4.12}$$

where $p_a(x)=1$ for $x\in(-a,a)$ and zero otherwise. Equivalently, the cross-ambiguity function, for $n\neq 0$, $m\neq 0$, is

$$A_{g_{\text{rx}},g_{\text{tx}}}(f,t) = \begin{cases} 0 & f\in(m\Delta f-\nu_{\max}, m\Delta f+\nu_{\max}) \;\; t\in(nT-\tau_{\max}, nT+\tau_{\max}), \\ 1 & f\in(-\nu_{\max},\nu_{\max}) \;\; t\in(-\tau_{\max},\tau_{\max}). \end{cases}$$

Although ideal pulses cannot be realized in practice, they can be approximated by waveforms with a support concentrated as much as possible in time and in frequency, given the constraint imposed by the uncertainty principle. Nonetheless, it is important to study the error performance of OTFS with ideal waveforms, since it can serve as a lower bound for performance of OTFS with practical waveforms, such as rectangular waveforms.

4.2.1 Time-frequency domain analysis

If g_{tx} and g_{rx} are perfectly localized in time and frequency, then they satisfy the biorthogonality condition. Omitting the noise term, we obtain

$$\mathbf{Y}_{\mathrm{tf}}[l,k] = \mathbf{H}_{\mathrm{tf}}[l,k]\mathbf{X}_{\mathrm{tf}}[l,k], \tag{4.13}$$

where $\mathbf{H}_{\mathrm{tf}} \in \mathbb{C}^{M\times N}$ is the *time-frequency domain channel matrix* with entries

$$\mathbf{H}_{\mathrm{tf}}[l,k] = \iint h(\tau,\nu)e^{j2\pi\nu kT}e^{-j2\pi l\Delta f\tau}d\tau d\nu. \tag{4.14}$$

for $l = 0,\ldots,M-1$ and $k = 0,\ldots,N-1$.

4.2.2 Delay-Doppler domain analysis

The delay-Doppler transmitted and received samples matrices are related to those in the time-frequency domain by SFFT:

$$\mathbf{Y} = \mathrm{SFFT}(\mathbf{Y}_{\mathrm{tf}}),$$
$$\mathbf{X} = \mathrm{SFFT}(\mathbf{X}_{\mathrm{tf}}).$$

The delay-Doppler domain channel matrix of size $M\times N$ is related to the time-frequency channel matrix as as

$$\mathbf{H}_{\mathrm{dd}} = \mathrm{SFFT}(\mathbf{H}_{\mathrm{tf}}),$$

where

$$\mathbf{H}_{\mathrm{dd}}[m,n] = \sum_l\sum_k \mathbf{H}_{\mathrm{tf}}[l,k]e^{-j2\pi(\frac{nk}{N}-\frac{ml}{M})}.$$

Using the SFFT property (see Appendix B.5), for pulses satisfying the biorthogonality condition in the time-frequency domain, the received signal in delay-Doppler domain can be written (by taking SFFT of the left-hand side and right-hand side of (4.13)) as

$$\begin{aligned}\mathbf{Y}[m,n] &= \mathbf{H}_{\mathrm{dd}}[m,n] \circledast \mathbf{X}[m,n] \qquad \text{(2D circular convolution)}\\ &= \sum_{m'}\sum_{n'}\mathbf{H}_{\mathrm{dd}}[m',n']\mathbf{X}[[m-m']_M],[n-n']_N]\\ &= \sum_{i=1}^{P} g_i\mathbf{X}\big[[m-l_i]_M,[n-k_i]_N\big]\end{aligned} \tag{4.15}$$

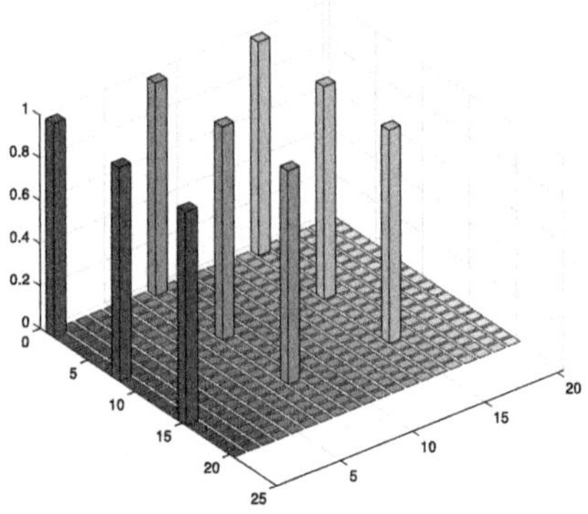

(a) 2D input signal $\mathbf{X}$

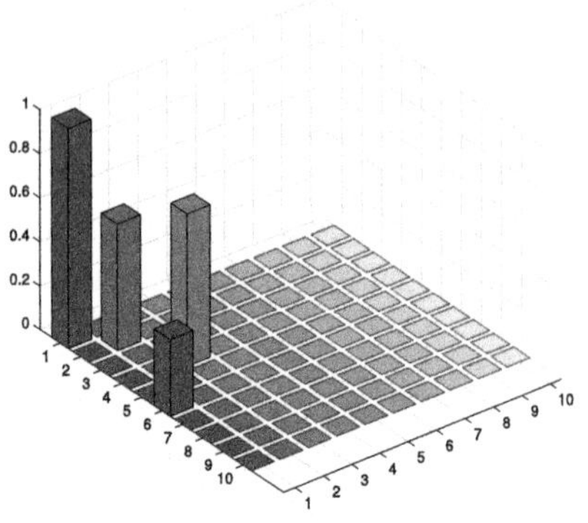

(b) delay-Doppler channel $\mathbf{H}_{\mathrm{dd}}$

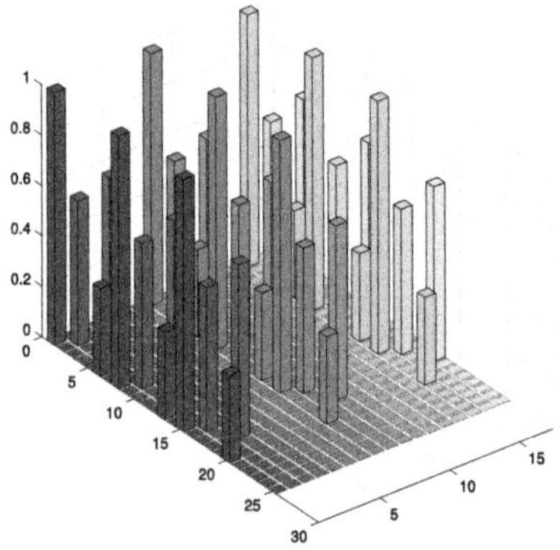

(c) 2D output signal $\mathbf{Y} = \mathbf{H}_{\mathrm{dd}} \circledast \mathbf{X}$

FIGURE 4.4 Received 2D signal (c) for ideal pulses as the 2D circular convolution of the input (a) and the channel (b).

for the case of integer delay and Doppler taps $\ell_i = l_i$ and $\kappa_i = k_i$. Fig. 4.4 shows the received signal as a 2D circular convolution of the transmitted signal $\mathbf{X}$ with the channel $\mathbf{H}$. Even though, as mentioned earlier, perfect biorthogonality cannot be achieved in practical cases, pulse shaping can be used to reduce the effect of the loss of biorthogonality. To quantify the loss of biorthogonality, one can measure the energy leakage outside each time-frequency resource block. Using narrow pulses as shown in Fig. 4.5 can help achieve approximate biorthogonality in the time-frequency domain at the cost of time-frequency resource efficiency. When approximate biorthogonality is achieved, one can use single tap equalization.

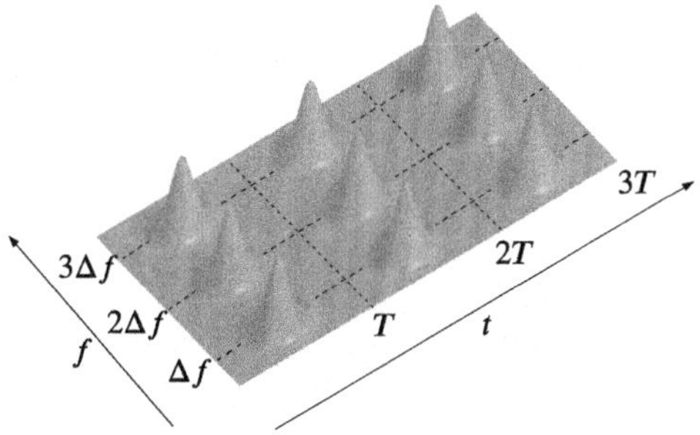

FIGURE 4.5 Time-frequency domain representation of transmitted signals with PS-OFDM.

4.3 Matrix formulation for OTFS

In this section, we present a more compact matrix formulation for OTFS, including OTFS modulation, demodulation, and the input–output relation in different domains.

4.3.1 OTFS modulation

Recall the information symbol matrix $\mathbf{X} \in \mathbb{C}^{M\times N}$ in the delay-Doppler domain, where each symbol is taken from a modulation alphabet $\mathbb{A} = \{a_1, \ldots, a_Q\}$ of size Q. Then, the time-frequency domain information samples matrix generated via ISFFT can be written as

$$\mathbf{X}_{\text{tf}} = \mathbf{F}_M \cdot \mathbf{X} \cdot \mathbf{F}_N^{\dagger} \tag{4.16}$$

where $\mathbf{F}_M$, $\mathbf{F}_N^{\dagger}$ represent the M-point Fourier transform and the N-point inverse Fourier transform, given in Appendix B.

The time-frequency domain matrix $\mathbf{X}_{\text{tf}} \in \mathbb{C}^{M\times N}$ is converted to the delay-time domain samples matrix $\tilde{\mathbf{X}} \in \mathbb{C}^{M\times N}$ via an M-point IFFT as

$$\tilde{\mathbf{X}} = \mathbf{F}_M^{\dagger} \cdot \mathbf{X}_{\text{tf}} = \underbrace{\mathbf{F}_M^{\dagger} \cdot \mathbf{F}_M}_{\mathbf{I}_M} \cdot \mathbf{X} \cdot \mathbf{F}_N^{\dagger} = \mathbf{X} \cdot \mathbf{F}_N^{\dagger}. \tag{4.17}$$

where $\mathbf{I}_M$ denotes the $M \times M$ identity matrix. The last step is equivalent to an *inverse discrete Zak transform* (IDZT). After the transmitter pulse shaping of the delay-time matrix $\tilde{\mathbf{X}}$, followed by the row-wise vectorization, we obtain the NM length time domain samples vector

$$\mathbf{s} = \text{vec}(\mathbf{G}_{\text{tx}} \cdot \tilde{\mathbf{X}}) \in \mathbb{C}^{NM\times 1}, \tag{4.18}$$

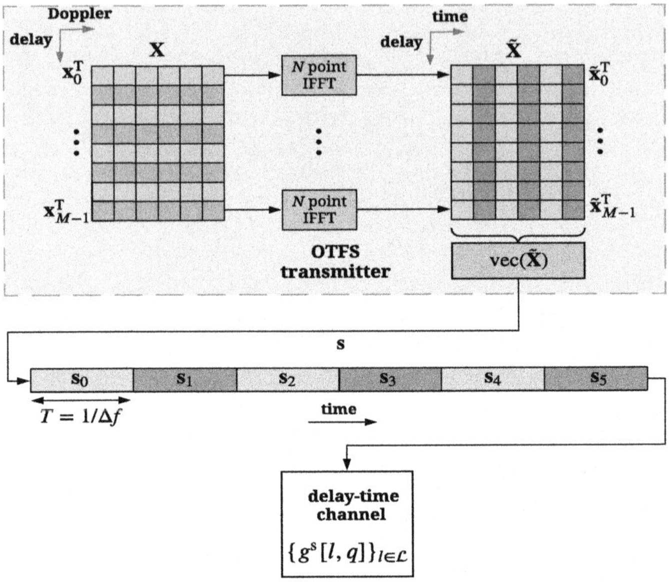

FIGURE 4.6 OTFS transmitter based on the IDZT using rectangular pulse shaping waveform ($M = 8$, $N = 6$).

where the diagonal matrix $\mathbf{G}_{\text{tx}}$ has the samples of $g_{\text{tx}}(t)$ as its entries,

$$\mathbf{G}_{\text{tx}} = \text{diag}\big[g_{\text{tx}}(0), g_{\text{tx}}(T/M), \ldots, g_{\text{tx}}\big((M-1)T/M\big)\big] \in \mathbb{C}^{M\times M},$$

and for rectangular waveforms, it reduces to an $M \times M$ identity matrix

$$\mathbf{G}_{\text{tx}} = \mathbf{I}_M.$$

The steps in (4.17) and (4.18) together constitute the Heisenberg transform. The time domain samples vector $\mathbf{s}$ are then DA converted and transmitted into the wireless medium as $s(t)$.

4.3.2 OTFS modulation via the IDZT

As shown in (4.17), OTFS modulation is equivalent to an IDZT, converting $\mathbf{X}$ in the delay-Doppler domain to $\tilde{\mathbf{X}}$ in the delay-time domain. Fig. 4.6 shows the equivalent transmitter operation of IDZT, assuming the rectangular pulse shaping waveform $\mathbf{G}_{\text{tx}} = \mathbf{I}_M$.

In the figure, the delay-Doppler and delay-time matrices are split into vectors $\mathbf{x}_m \in \mathbb{C}^{N\times 1}$ and $\tilde{\mathbf{x}}_m \in \mathbb{C}^{N\times 1}$, respectively. Taking transpose of (4.17) yields

$$\tilde{\mathbf{X}}^{\text{T}} = [\tilde{\mathbf{x}}_0, \ldots, \tilde{\mathbf{x}}_{M-1}] = \mathbf{F}_N^{\dagger}[\mathbf{x}_0, \ldots, \mathbf{x}_{M-1}] = \mathbf{F}_N^{\dagger} \cdot \mathbf{X}^{\text{T}}, \tag{4.19}$$

where $(\mathbf{F}_N^{\dagger})^T = \mathbf{F}_N^{\dagger}$ due to the symmetry of $\mathbf{F}_N^{\dagger}$. Then after parallel-to-serial conversion, as $\mathbf{G}_{\text{tx}} = \mathbf{I}_M$, the time domain samples in (4.18) reduce to

$$\mathbf{s} = \text{vec}(\tilde{\mathbf{X}}) = \begin{bmatrix} \mathbf{s}_0 \\ \vdots \\ \mathbf{s}_{N-1} \end{bmatrix} = \{\mathbf{s}[q],\ q = 0, \ldots, NM-1\}, \tag{4.20}$$

where each block $\mathbf{s}_n \in \mathbb{C}^{M\times 1}, n = 0, \ldots, N-1$, has M time domain samples, which occupy $T = 1/\Delta f$ seconds.

From Fig. 4.6, we can directly relate $\mathbf{X}$ and $\mathbf{s}$ from (4.19) and (4.20) as

$$\boxed{\mathbf{s} = \text{vec}(\tilde{\mathbf{X}}) = \text{vec}(\mathbf{X} \cdot \mathbf{F}_N^{\dagger})}, \tag{4.21}$$

with entries

$$\boxed{\mathbf{s}[q] = \mathbf{s}[m + nM] = \frac{1}{\sqrt{N}} \sum_{p=0}^{N-1} \mathbf{X}[m,\, p] e^{j2\pi pn/N},} \tag{4.22}$$

where $q = m + nM$ for $m = 0, \ldots, M-1$ and $n = 0, \ldots, N-1$, which is exactly the IDZT of $\mathbf{X}$ (see Chapter 5 about the Zak transform and note that the delay and Doppler indices used there are l, k following the Zak transform conventional notations, different from m, n used in this chapter).

MATLAB® implementation of the OTFS transmitter is given in MATLAB codes 2 and 3 of Appendix C.

4.3.3 OTFS demodulation

At the receiver, we obtain the received NM complex samples vector as

$$\mathbf{r} = \begin{bmatrix} \mathbf{r}_0 \\ \vdots \\ \mathbf{r}_{N-1} \end{bmatrix} = \{\mathbf{r}[q], q = 0, \ldots, NM-1\}, \tag{4.23}$$

where $\mathbf{r}_n \in \mathbb{C}^{M\times 1}$, for $n = 0, \ldots, N-1$. The vector $\mathbf{r}$ is converted to the delay-time matrix $\tilde{\mathbf{Y}} \in \mathbb{C}^{M\times N}$ as

$$\tilde{\mathbf{Y}} = \mathbf{G}_{\text{rx}} \cdot \left(\text{vec}_{M,N}^{-1}(\mathbf{r})\right), \tag{4.24}$$

where the operator $\text{vec}_{M,N}^{-1}(\mathbf{r})$ converts $\mathbf{r} \in \mathbb{C}^{NM\times 1}$ in (4.23) to a $M \times N$ matrix, and the diagonal matrix $\mathbf{G}_{\text{rx}}$ is the receiver pulse shaping matrix, defined as

$$\mathbf{G}_{\text{rx}} = \text{diag}\left[g_{\text{rx}}(0), g_{\text{rx}}(T/M), \ldots, g_{\text{rx}}\left((M-1)T/M\right)\right] \in \mathbb{C}^{M\times M}.$$

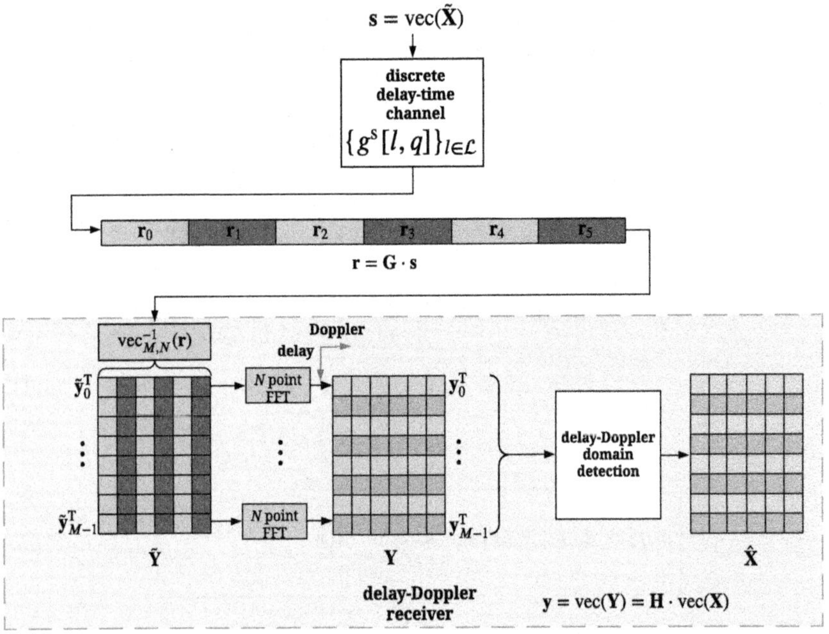

FIGURE 4.7 OTFS receiver based on the DZT using rectangular pulse shaping waveform ($M = 8$, $N = 6$).

For rectangular pulse shaping waveforms $\mathbf{G}_{\text{rx}} = \mathbf{I}_M$, (4.24) reduces to

$$\tilde{\mathbf{Y}} = \text{vec}^{-1}_{M,N}(\mathbf{r}) \;\text{ or }\; \mathbf{r} = \text{vec}(\tilde{\mathbf{Y}}). \tag{4.25}$$

Then the time-frequency domain received samples matrix $\mathbf{Y}_{\text{tf}} \in \mathbb{C}^{M\times N}$ is obtained by performing an M-point DFT operation on the delay-time samples

$$\mathbf{Y}_{\text{tf}} = \mathbf{F}_M \cdot \tilde{\mathbf{Y}}. \tag{4.26}$$

The operations in (4.24) and (4.26) constitute the Wigner transform. An SFFT operation is performed to get back the delay-Doppler domain symbols. The SFFT operation in cascade with the Wigner transform can be simplified as

$$\mathbf{Y} = \mathbf{F}_M^{\dagger} \cdot \mathbf{Y}_{\text{tf}} \cdot \mathbf{F}_N = \tilde{\mathbf{Y}} \cdot \mathbf{F}_N. \tag{4.27}$$

4.3.4 OTFS demodulation via the DZT

As shown in (4.24) and (4.27), OTFS demodulation is equivalent to a DZT, converting the received samples in the time domain to the delay-

Doppler domain. Fig. 4.7 shows the equivalent receiver operation of DZT for rectangular pulse shaping waveform $\mathbf{G}_{\text{rx}} = \mathbf{I}_M$. In the figure, the delay-Doppler and delay-time matrices $\mathbf{Y}$ and $\tilde{\mathbf{Y}}$ are split into vectors $\mathbf{y}_m, \tilde{\mathbf{y}}_m \in \mathbb{C}^{N\times 1}$ for $m = 0, \ldots, M-1$. Taking the transpose of (4.27) yields

$$\mathbf{Y}^{\text{T}} = [\mathbf{y}_0, \ldots, \mathbf{y}_{M-1}] = \mathbf{F}_N[\tilde{\mathbf{y}}_0, \ldots, \tilde{\mathbf{y}}_{M-1}] = \mathbf{F}_N \cdot \tilde{\mathbf{Y}}^{\text{T}}, \tag{4.28}$$

where $(\mathbf{F}_N)^T = \mathbf{F}_N$ due to the symmetry of $\mathbf{F}_N$.

From Fig. 4.7, we can directly relate $\mathbf{r}$ and $\mathbf{Y}$ via (4.24) and (4.28) as

$$\boxed{\mathbf{Y} = \tilde{\mathbf{Y}} \cdot \mathbf{F}_N = (\text{vec}_{M,N}^{-1}(\mathbf{r})) \cdot \mathbf{F}_N,} \tag{4.29}$$

with entries

$$\boxed{\mathbf{Y}[m,n] = \frac{1}{\sqrt{N}} \sum_{p=0}^{N-1} \mathbf{r}[m+pM] e^{-j2\pi np/N},} \tag{4.30}$$

for $m = 0, \ldots M-1$ and $n = 0, \ldots N-1$, which is the DZT of $\mathbf{r}$, i.e., $\mathbf{Y} = Z_M[\mathbf{r}]$ (see Chapter 5 and note that the delay and Doppler indices used there are l, k following the Zak transform conventional notations, different from m, n used in this chapter).

MATLAB implementation of the OTFS receiver is given in MATLAB code 12 of Appendix C.

4.4 OTFS input–output relations in vectorized form

In this section, we present the OTFS input–output relation in vectorized form for practical rectangular pulse shaping waveforms ($\mathbf{G}_{\text{tx}} = \mathbf{G}_{\text{rx}} = \mathbf{I}_M$). Our analysis will be conducted in different domains including time, time-frequency, delay-time, and delay-Doppler.

We first introduce the relevant notations. In the delay-Doppler domain, vectorizing $\mathbf{X}^T$, $\mathbf{Y}^T$ in (4.19) and (4.28) yields the transmitted and received symbol vectors $\mathbf{x} \in \mathbb{C}^{NM\times 1}$ and $\mathbf{y} \in \mathbb{C}^{NM\times 1}$, each having M blocks of N samples, i.e.,

$$\mathbf{x} = \begin{bmatrix} \mathbf{x}_0 \\ \vdots \\ \mathbf{x}_{M-1} \end{bmatrix} = \text{vec}(\mathbf{X}^{\text{T}}), \qquad \mathbf{y} = \begin{bmatrix} \mathbf{y}_0 \\ \vdots \\ \mathbf{y}_{M-1} \end{bmatrix} = \text{vec}(\mathbf{Y}^{\text{T}}), \tag{4.31}$$

where the subvectors $\mathbf{x}_m, \mathbf{y}_m \in \mathbb{C}^{N\times 1}$ for $m = 0, \ldots, M-1$ are along the Doppler axis, as shown in Figs. 4.6 and 4.7.

Next, in the time domain, we recall the transmitted and received samples vectors in (4.20) and (4.23), each having N blocks of M samples as

$$\mathbf{s} = \begin{bmatrix} \mathbf{s}_0 \\ \vdots \\ \mathbf{s}_{N-1} \end{bmatrix} \qquad \mathbf{r} = \begin{bmatrix} \mathbf{r}_0 \\ \vdots \\ \mathbf{r}_{N-1} \end{bmatrix}$$

where $\mathbf{s}_n,\ \mathbf{r}_n \in \mathbb{C}^{M\times 1}$ are the transmitted and received samples subvectors for $n = 0, \dots, N-1$. They are related to the delay-Doppler domain vectors $\mathbf{x}$ and $\mathbf{y}$ as

$$\boxed{\mathbf{s} = \mathbf{P} \cdot \left(\mathbf{G}_{\text{tx}} \otimes \mathbf{F}_N^{\dagger}\right) \cdot \mathbf{x},}$$
$$\boxed{\mathbf{r} = \mathbf{P} \cdot \left(\mathbf{G}_{\text{rx}} \otimes \mathbf{F}_N^{\dagger}\right) \cdot \mathbf{y},} \tag{4.32}$$

where $\mathbf{P} \in \mathbb{Z}^{NM\times NM}$ is the *row-column interleaver* matrix due to vectorizing of $\mathbf{X}^{\mathrm{T}}$ and $\mathbf{Y}^{\mathrm{T}}$ [1]. The permutation operation by $\mathbf{P}$ on any NM length vector $\mathbf{a}$ can be considered as writing the elements of $\mathbf{a}$ into an $N \times M$ matrix $\mathbf{A}$ column-wise and then reading out row-wise. The $NM \times NM$ matrix $\mathbf{P}$ can be written as

$$\mathbf{P} = \begin{bmatrix} \mathbf{E}_{1,1} & \mathbf{E}_{2,1} & \cdots & \mathbf{E}_{M,1} \\ \mathbf{E}_{1,2} & \mathbf{E}_{2,2} & \cdots & \mathbf{E}_{M,2} \\ \vdots & \ddots & \ddots & \vdots \\ \mathbf{E}_{1,N} & \mathbf{E}_{2,N} & \cdots & \mathbf{E}_{M,N} \end{bmatrix}, \tag{4.33}$$

where the $M \times N$ matrix $\mathbf{E}_{i,j}$ is defined as

$$\mathbf{E}_{i,j}[i', j'] = \begin{cases} 1, & \text{if } i' = i \text{ and } j' = j, \\ 0, & \text{otherwise,} \end{cases} \tag{4.34}$$

for $i, i' \in [1, M]$ and $j, j' \in [1, N]$. The Matlab implementation of the permutation matrix $\mathbf{P}$ can be found in MATLAB code 3 of Appendix C.

Assuming rectangular pulses for the rest of the chapter,

$$\mathbf{G}_{\text{tx}} = \mathbf{G}_{\text{rx}} = \mathbf{I}_M,$$

the modulation and demodulation steps in (4.32) can be rewritten as

$$\boxed{\mathbf{s} = \mathbf{P} \cdot \left(\mathbf{I}_M \otimes \mathbf{F}_N^{\dagger}\right) \cdot \mathbf{x},}$$
$$\boxed{\mathbf{r} = \mathbf{P} \cdot \left(\mathbf{I}_M \otimes \mathbf{F}_N^{\dagger}\right) \cdot \mathbf{y}.} \tag{4.35}$$

[1] This is different from [5], where vectorizations of $\mathbf{X}$ and $\mathbf{Y}$ are performed to obtain symbol vectors along the delay axis, and thus $\mathbf{P}$ is not needed in (4.32).

In Table 4.1 we summarize the notations used in the rest of the book to identify channel matrices in the different domains: $\tilde{\cdot}$ for delay-time and $\check{\cdot}$ for time-frequency. The same convention is used for the vectors.

TABLE 4.1 Summary of channel matrices notations.

$\mathbf{G}$	$NM \times NM$	*time domain channel matrix* (Fig. 4.8)
$\mathbf{G}_{0,0},\ldots,\mathbf{G}_{N-1,0}$	$M \times M$	diagonal subblocks of $\mathbf{G}$
$\mathbf{G}_{1,1},\ldots,\mathbf{G}_{N-1,1}$	$M \times M$	1st subdiagonal subblocks of $\mathbf{G}$
$\mathbf{G}_{0,1}$	$M \times M$	top-right corner subblock of $\mathbf{G}$ (only in RCP-OTFS)
$\check{\mathbf{H}}$	$NM \times NM$	*time-frequency channel matrix* (Fig. 4.9)
$\check{\mathbf{H}}_{0,0},\ldots,\check{\mathbf{H}}_{N-1,0}$	$M \times M$	diagonal subblocks of $\check{\mathbf{H}}$
$\check{\mathbf{H}}_{1,1},\ldots,\check{\mathbf{H}}_{N-1,1}$	$M \times M$	1st subdiagonal subblocks of $\check{\mathbf{H}}$
$\tilde{\mathbf{H}}$	$NM \times NM$	*delay-time channel matrix* (Fig. 4.11)
$\tilde{\mathbf{K}}_{m,l}$	$N \times N$	$(m,[m-l]_M)$-th subblock of $\tilde{\mathbf{H}}$
$\tilde{\boldsymbol{\nu}}_{m,l}$	$N \times 1$	delay-time channel vectors (diagonal elements of $\tilde{\mathbf{K}}_{m,l}$)
$\mathbf{H}$	$NM \times NM$	*delay-Doppler channel matrix* (Fig. 4.12)
$\mathbf{K}_{m,l}$	$N \times N$	$(m,[m-l]_M)$-th subblock of $\mathbf{H}$
$\boldsymbol{\nu}_{m,l}$	$N \times 1$	*delay-Doppler channel vectors* (1st column of $\mathbf{K}_{m,l}$)

4.4.1 Time domain input–output relation

Considering the discrete-time input–output relation using the vector notation, we can rewrite (4.5) with the noise term as

$$\mathbf{r}[q] = \sum_{l\in\mathcal{L}} g^{\mathrm{s}}[l,q]\mathbf{s}[q-l] + \mathbf{w}[q], \quad q = 0,\ldots,NM-1, \tag{4.36}$$

where we assume $\mathbf{s}[q-l]=0$ for $q-l<0$, $g^{\mathrm{s}}[l,q]$ is the time-variant impulse response $g(\tau,t)$ sampled at $\tau = lT/M$ and $t = qT/M$, and $\mathbf{w}[q]$ is the independent and identically distributed additive white Gaussian noise (AWGN) with variance σ_w^2.

The vectorized input–output relation in (4.36) can be written as

$$\boxed{\mathbf{r} = \mathbf{G}\cdot\mathbf{s} + \mathbf{w},} \tag{4.37}$$

where $\mathbf{w} \in \mathbb{C}^{NM\times 1}$ is the time domain AWGN vector, $\mathbf{s}$ is the transmitted samples vector given in (4.20), and $\mathbf{G} \in \mathbb{C}^{NM\times NM}$ is the channel matrix.

Next, we derive the *block-wise* input–output relation of (4.37). We start with the matrix $\mathbf{G}$, which has $l_{\max}+1$ *nonzero diagonal and subdiagonals,* with elements in each subdiagonal given by

$$\mathbf{G}[q,q-l] = g^{\mathrm{s}}[l,q] \quad \text{for } l \le q \le NM-1 \tag{4.38}$$

and zero elsewhere. For example, when $l=0$, we obtain the diagonal elements $\mathbf{G}[q,q]$ for all q. When $l=1$, we obtain the first subdiagonal

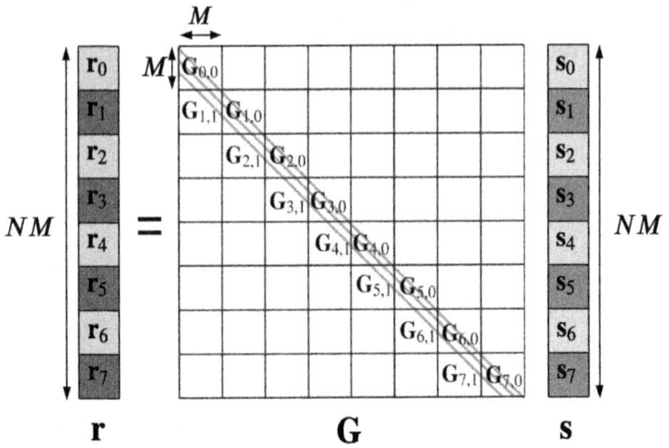

FIGURE 4.8 Time domain channel matrix **G** with three delay paths represented by the three green (light gray in print version) subdiagonals partitioned into $M \times M$ submatrices.

elements $\mathbf{G}[q, q-1]$ for all $q \geq 1$, and so on. Fig. 4.8 shows an example describing the channel matrix $\mathbf{G}$ with three delay paths ($l_{\max} = 2$) and $N = 8$, where $\mathbf{s}_n \in \mathbb{C}^{M \times 1}$ and $\mathbf{r}_n \in \mathbb{C}^{M \times 1}$, $n = 0, \ldots, N-1$, denote the transmitted and received time domain vectors.

Then the block-wise input–output relation can be written as

$$\begin{cases} \mathbf{r}_0 = \mathbf{G}_{0,0} \cdot \mathbf{s}_0 + \mathbf{w}_0, & n = 0, \\ \mathbf{r}_n = \mathbf{G}_{n,0} \cdot \mathbf{s}_n + \mathbf{G}_{n,1} \cdot \mathbf{s}_{n-1} + \mathbf{w}_n, & 1 \leq n \leq N-1, \end{cases} \tag{4.39}$$

where $\mathbf{w}_n \in \mathbb{C}^{M \times 1}$ is the n-th block AWGN vector and $\mathbf{G}_{n,n'} \in \mathbb{C}^{M \times M}$ represents the channel between the n-th received and $(n - n')$-th transmitted blocks. Since $l_{\max} < M$, the n-th received block can have interference only from the $(n-1)$-th block (see 4.39), which implies $n' = 0$ or 1. When $n' = 0$, $\mathbf{G}_{n,0}$, $n = 0, \ldots, N-1$, are the *diagonal blocks* of $\mathbf{G}$, while for $n' = 1$, $\mathbf{G}_{n,1}$, $n > 0$, are the *first subdiagonal blocks* of $\mathbf{G}$, each representing the interblock interference between the n-th received and the $(n-1)$-th transmitted blocks.

4.4.2 Time-frequency input–output relation

The transmitted and received symbol blocks $\check{\mathbf{x}}, \check{\mathbf{y}} \in \mathbb{C}^{NM \times 1}$ in the time-frequency domain are obtained by taking the M-point DFT of the time domain blocks $\mathbf{s}, \mathbf{r} \in \mathbb{C}^{NM \times 1}$ as

$$\check{\mathbf{x}} = (\mathbf{I}_N \otimes \mathbf{F}_M) \cdot \mathbf{s},$$
$$\check{\mathbf{y}} = (\mathbf{I}_N \otimes \mathbf{F}_M) \cdot \mathbf{r}. \tag{4.40}$$

Multiplying $(\mathbf{I}_N \otimes \mathbf{F}_M)$ in both sides the time domain relation

$$\mathbf{r} = \mathbf{G} \cdot \mathbf{s} + \mathbf{w}$$

yields

$$\begin{aligned}
\underbrace{(\mathbf{I}_N \otimes \mathbf{F}_M)\mathbf{r}}_{\check{\mathbf{y}}} &= (\mathbf{I}_N \otimes \mathbf{F}_M)\mathbf{G} \cdot \mathbf{s} + (\mathbf{I}_N \otimes \mathbf{F}_M)\mathbf{w} \\
&= (\mathbf{I}_N \otimes \mathbf{F}_M)\mathbf{G}(\mathbf{I}_N \otimes \mathbf{F}_M)^{\dagger}(\mathbf{I}_N \otimes \mathbf{F}_M) \cdot \mathbf{s} + (\mathbf{I}_N \otimes \mathbf{F}_M)\mathbf{w} \\
&= \underbrace{(\mathbf{I}_N \otimes \mathbf{F}_M)\mathbf{G}\big(\mathbf{I}_N \otimes \mathbf{F}_M^{\dagger}\big)}_{\check{\mathbf{H}}} \underbrace{(\mathbf{I}_N \otimes \mathbf{F}_M) \cdot \mathbf{s}}_{\check{\mathbf{x}}} + \underbrace{(\mathbf{I}_N \otimes \mathbf{F}_M)\mathbf{w}}_{\check{\mathbf{w}}},
\end{aligned}$$

where the second step is due to $(\mathbf{I}_N \otimes \mathbf{F}_M)^{\dagger}(\mathbf{I}_N \otimes \mathbf{F}_M) = \mathbf{I}_{NM}$ and the last step is due to $(\mathbf{I}_N \otimes \mathbf{F}_M)^{\dagger} = (\mathbf{I}_N \otimes \mathbf{F}_M^{\dagger})$ with symmetric $\mathbf{F}_M$. Hence, the input–output relation in the time-frequency domain is

$$\boxed{\check{\mathbf{y}} = \check{\mathbf{H}} \cdot \check{\mathbf{x}} + \check{\mathbf{w}}} \tag{4.41}$$

with the time-frequency domain channel matrix and AWGN vector

$$\check{\mathbf{H}} = (\mathbf{I}_N \otimes \mathbf{F}_M) \cdot \mathbf{G} \cdot \big(\mathbf{I}_N \otimes \mathbf{F}_M^{\dagger}\big), \tag{4.42}$$

$$\check{\mathbf{w}} = (\mathbf{I}_N \otimes \mathbf{F}_M) \cdot \mathbf{w}. \tag{4.43}$$

Next, we derive the block-wise and element-wise input–output relation of (4.41), respectively. Let the time-frequency samples be split into N blocks as

$$\check{\mathbf{x}} = \begin{bmatrix} \check{\mathbf{x}}_0 \\ \vdots \\ \check{\mathbf{x}}_{N-1} \end{bmatrix}, \quad \check{\mathbf{y}} = \begin{bmatrix} \check{\mathbf{y}}_0 \\ \vdots \\ \check{\mathbf{y}}_{N-1} \end{bmatrix}, \tag{4.44}$$

where $\check{\mathbf{x}}_n, \check{\mathbf{y}}_n \in \mathbb{C}^{M \times 1}$ for $n = 0, \ldots, N-1$, are related to the time domain blocks of length M as

$$\check{\mathbf{x}}_n = \mathbf{F}_M \cdot \mathbf{s}_n, \quad \check{\mathbf{y}}_n = \mathbf{F}_M \cdot \mathbf{r}_n. \tag{4.45}$$

From (4.39), (4.40), and (4.45), the time-frequency block-wise input–output relation can be obtained as

$$\boxed{\begin{cases} \check{\mathbf{y}}_0 = \check{\mathbf{H}}_{0,0} \cdot \check{\mathbf{x}}_0 + \check{\mathbf{w}}_0, & n = 0, \\ \check{\mathbf{y}}_n = \check{\mathbf{H}}_{n,0} \cdot \check{\mathbf{x}}_n + \check{\mathbf{H}}_{n,1} \cdot \check{\mathbf{x}}_{n-1} + \check{\mathbf{w}}_n, & 1 \le n \le N-1, \end{cases}} \tag{4.46}$$

where, for rectangular waveforms,

$$\check{\mathbf{H}}_{n,n'} = \mathbf{F}_M \cdot \mathbf{G}_{n,n'} \cdot \mathbf{F}_M^{\dagger} \tag{4.47}$$

is the time-frequency domain channel between the n-th received block and the $(n-n')$-th transmitted block with $n' \in \{0, 1\}$. Fig. 4.9 shows the structure of the time-frequency channel matrix $\check{\mathbf{H}}$ with $N = 8$. Similar to the time domain channel matrix structure, $\check{\mathbf{H}}_{n,0}$ and $\check{\mathbf{H}}_{n,1}$, $n = 0, \ldots, N-1$, are the diagonal blocks and the first subdiagonal blocks of $\check{\mathbf{H}}$, respectively, where $\check{\mathbf{H}}_{n,1}$ represents the interblock interference between the n-th received and the $(n-1)$-th transmitted blocks.

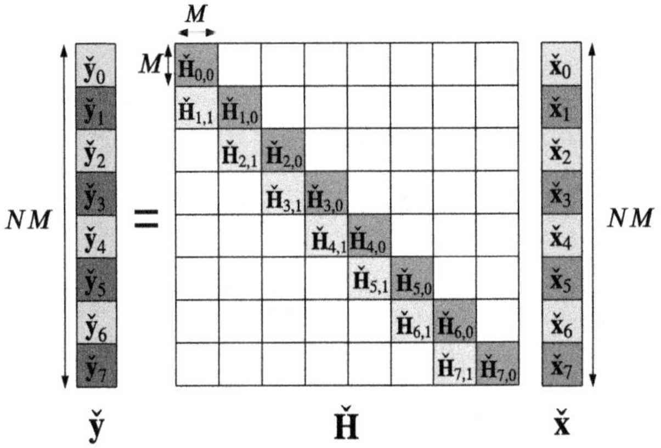

FIGURE 4.9 Time-frequency channel matrix $\check{\mathbf{H}}$ partitioned into $M \times M$ submatrices.

Omitting noise, the *element-wise* input–output relation in the time-frequency domain can be written as

$$\begin{aligned}\check{\mathbf{y}}_n[m] = {} & \check{\mathbf{H}}_{n,0}[m,m]\check{\mathbf{x}}_n[m] \\ & + \underbrace{\sum_{m'=0, m' \neq m}^{M-1} \check{\mathbf{H}}_{n,0}[m,m']\check{\mathbf{x}}_n[m']}_{\text{ICI}} + \underbrace{\sum_{m'=0}^{M-1} \check{\mathbf{H}}_{n,1}[m,m']\check{\mathbf{x}}_{n-1}[m]}_{\text{ISI}}\end{aligned} \tag{4.48}$$

for $m = 0, \ldots, M-1$. As shown in (4.48), the interference between the samples in the time-frequency domain can be broadly divided into ICI and ISI, due to channel Doppler spread and delay spread. As explained in Section 4.2, for the ideal pulse shaping waveform, the channel introduces no ISI and ICI between the time-frequency samples, i.e.,

$$\check{\mathbf{H}}_{n,0}[m,m'] = 0, \quad \text{for } m \neq m', \quad \check{\mathbf{H}}_{n,1}[m,m'] = 0, \quad \text{for all } m,\ m'. \tag{4.49}$$

The first condition implies that $\check{\mathbf{H}}_{n,0}$ must be a diagonal matrix. The second condition in (4.49) implies that there should be no ISI between blocks.

Now let us take a look at why the ideal pulse shaping waveform assumption is invalid for all practical channels. We start with a general time-frequency channel matrix with arbitrary pulse shaping waveforms $\mathbf{G}_{\text{tx}}$ and $\mathbf{G}_{\text{rx}}$, given by

$$\check{\mathbf{H}}_{n,n'} = \mathbf{F}_M \cdot \underbrace{\mathbf{G}_{\text{rx}} \cdot \mathbf{G}_{n,n'} \cdot \mathbf{G}_{\text{tx}}}_{\mathbf{G}_{\text{eq}}} \cdot \mathbf{F}_M^{\dagger}. \tag{4.50}$$

where $\mathbf{G}_{\text{eq}}$ is the equivalent time domain channel matrix. Since $\mathbf{G}_{\text{tx}}$ and $\mathbf{G}_{\text{rx}}$ are diagonal matrices, $\check{\mathbf{H}}_{n,n'}$ can be diagonalized through Fourier transformations only if $\mathbf{G}_{\text{eq}}$ is a circulant matrix. However, the strcuture of $\mathbf{G}_{\text{eq}}$ is circulant only for static multipath channels, as seen for OFDM in Chapter 3. This means that no practical pulse shaping waveforms can make $\mathbf{G}_{\text{eq}}$ circulant.

4.4.3 Delay-time input–output relation

By vectorizing the transposed delay-time matrices $\tilde{\mathbf{X}}^{\text{T}}$ in (4.17) and $\tilde{\mathbf{Y}}^{\text{T}}$ in (4.24) yields the vectors $\tilde{\mathbf{x}}$ and $\tilde{\mathbf{y}}$, each containing M blocks of N samples, i.e.,

$$\tilde{\mathbf{x}} = \begin{bmatrix} \tilde{\mathbf{x}}_0 \\ \vdots \\ \tilde{\mathbf{x}}_{M-1} \end{bmatrix} = \text{vec}(\tilde{\mathbf{X}}^{\text{T}}), \quad \tilde{\mathbf{y}} = \begin{bmatrix} \tilde{\mathbf{y}}_0 \\ \vdots \\ \tilde{\mathbf{y}}_{M-1} \end{bmatrix} = \text{vec}(\tilde{\mathbf{Y}}^{\text{T}}), \tag{4.51}$$

where $\tilde{\mathbf{x}}_m, \tilde{\mathbf{y}}_m \in \mathbb{C}^{N\times 1}$ are subvectors of $\tilde{\mathbf{x}}, \tilde{\mathbf{y}}$, for $m = 0, \ldots, M-1$. The delay-time vectors are related to the time domain samples by the row-column permutation operation

$$\tilde{\mathbf{x}} = \mathbf{P}^{\text{T}} \cdot \mathbf{s}, \quad \tilde{\mathbf{y}} = \mathbf{P}^{\text{T}} \cdot \mathbf{r}. \tag{4.52}$$

The row-column interleaving operation is illustrated in Fig. 4.10. Readers should note that the delay-time samples are split into M vectors $\tilde{\mathbf{x}}_m$ of length N, whereas the time domain samples are split into N vectors $\mathbf{s}_n$ of length M.

Multiplying both sides of the time domain input–output relation $\mathbf{r} = \mathbf{G} \cdot \mathbf{s} + \mathbf{w}$ with $\mathbf{P}^{\text{T}}$ yields

$$\underbrace{\mathbf{P}^{\text{T}} \cdot \mathbf{r}}_{\tilde{\mathbf{y}}} = \underbrace{\mathbf{P}^{\text{T}} \cdot \mathbf{G} \cdot \mathbf{P}}_{\tilde{\mathbf{H}}} \underbrace{\mathbf{P}^{\text{T}} \cdot \mathbf{s}}_{\tilde{\mathbf{x}}} + \underbrace{\mathbf{P}^{\text{T}} \cdot \mathbf{w}}_{\tilde{\mathbf{w}}}, \tag{4.53}$$

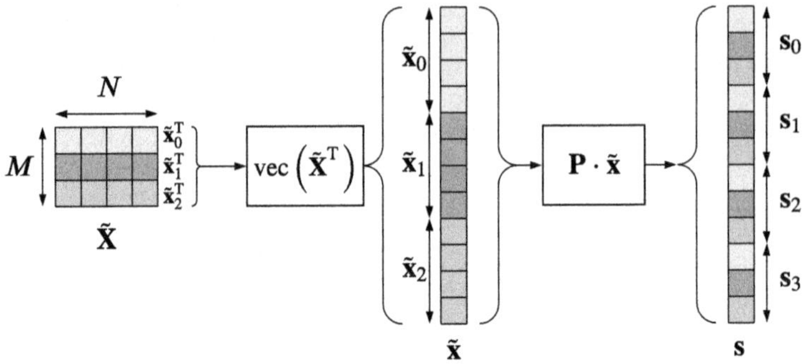

FIGURE 4.10 The row-column interleaving operation to generate the time domain samples from the delay-time samples using the permutation matrix **P** for an OTFS frame of size 12 with $M = 3$ and $N = 4$.

where $\mathbf{P}^{\mathrm{T}} \cdot \mathbf{P} = \mathbf{I}_{NM}$. We rewrite the delay-time input–output relation as

$$\boxed{\tilde{\mathbf{y}} = \tilde{\mathbf{H}} \cdot \tilde{\mathbf{x}} + \tilde{\mathbf{w}}} \tag{4.54}$$

with

$$\tilde{\mathbf{H}} = \mathbf{P}^{\mathrm{T}} \cdot \mathbf{G} \cdot \mathbf{P}, \tag{4.55}$$

$$\tilde{\mathbf{w}} = \mathbf{P}^{\mathrm{T}} \cdot \mathbf{w}. \tag{4.56}$$

Next, we derive the block-wise input–output relation of (4.54). Let $\tilde{\mathbf{K}}_{m,l} \in \mathbb{C}^{N\times N}$ be the subblocks of channel matrix $\tilde{\mathbf{H}}$, representing the delay-time channel block between the m-th received block and the $[m-l]_M$-th transmitted block, where $m = 0, \ldots, M-1$ and $l = 0, \ldots, l_{\max}-1$. Fig. 4.11 illustrates the delay-time channel matrix with $l_{\max} = 2$ and $M = 8$. Then the block-wise input–output relation in the delay-time domain can be written as

$$\boxed{\tilde{\mathbf{y}}_m = \sum_{l\in\mathcal{L}} \tilde{\mathbf{K}}_{m,l} \cdot \tilde{\mathbf{x}}_{[m-l]_M} + \tilde{\mathbf{w}}_m,} \tag{4.57}$$

where $m = 0, \ldots, M-1$ and $\tilde{\mathbf{w}}_m \in \mathbb{C}^{N\times 1}$ is the delay-time AWGN vector. Note that $[m-l]_M$ is due to the effect of the row-column interleaver and the interblock interference.

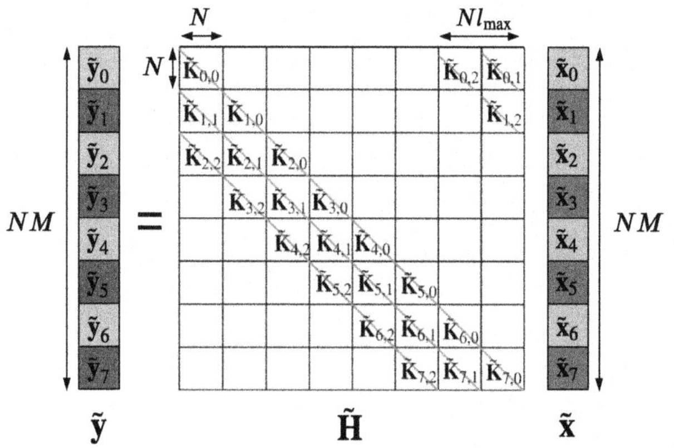

FIGURE 4.11 Delay-time domain channel matrix $\tilde{\mathbf{H}} = \mathbf{P}^{\mathrm{T}} \cdot \mathbf{G} \cdot \mathbf{P}$ with three delay paths partitioned into $N \times N$ submatrices.

4.4.4 Delay-Doppler input–output relation

Since $(\mathbf{P} \cdot (\mathbf{I}_M \otimes \mathbf{F}_N^{\dagger}))^{\dagger} = (\mathbf{I}_M \otimes \mathbf{F}_N)\mathbf{P}^T$, using the relations in (4.35) and multiplying both sides of $\mathbf{r} = \mathbf{G} \cdot \mathbf{s} + \mathbf{w}$ with $\mathbf{A} = (\mathbf{I}_M \otimes \mathbf{F}_N)\mathbf{P}^T$ yields

$$\underbrace{\mathbf{A} \cdot \mathbf{r}}_{\mathbf{y}} = \underbrace{\mathbf{A} \cdot \mathbf{G}\mathbf{A}^{\dagger}}_{\mathbf{H}} \underbrace{\mathbf{A} \cdot \mathbf{s}}_{\mathbf{x}} + \underbrace{\mathbf{A}\mathbf{w}}_{\mathbf{z}}, \tag{4.58}$$

where $\mathbf{A}^{\dagger}\mathbf{A} = \mathbf{I}_{NM \times NM}$. This leads to the delay-Doppler input–output equation

$$\boxed{\mathbf{y} = \mathbf{H} \cdot \mathbf{x} + \mathbf{z},} \tag{4.59}$$

where the $NM \times NM$ delay-Doppler channel matrix and the $NM \times 1$ AWGN vector are

$$\begin{aligned} \mathbf{H} &= (\mathbf{I}_M \otimes \mathbf{F}_N) \cdot \left(\mathbf{P}^{\mathrm{T}} \cdot \mathbf{G} \cdot \mathbf{P}\right) \cdot \left(\mathbf{I}_M \otimes \mathbf{F}_N^{\dagger}\right) \\ &= (\mathbf{I}_M \otimes \mathbf{F}_N) \cdot \tilde{\mathbf{H}} \cdot \left(\mathbf{I}_M \otimes \mathbf{F}_N^{\dagger}\right), \end{aligned} \tag{4.60}$$

$$\begin{aligned} \mathbf{z} &= (\mathbf{I}_M \otimes \mathbf{F}_N) \cdot \left(\mathbf{P}^{\mathrm{T}} \cdot \mathbf{w}\right) \\ &= (\mathbf{I}_M \otimes \mathbf{F}_N) \cdot \tilde{\mathbf{w}}. \end{aligned} \tag{4.61}$$

Next, we focus on the block-wise delay-Doppler input–output relation of OTFS. Let $\mathbf{K}_{m,l} \in \mathbb{C}^{N \times N}$ be the subblocks of channel matrix $\mathbf{H}$, representing the delay-Doppler channel block between the m received block and the $[m-l]_M$-th transmitted block, where $m = 0, \ldots, M-1, l = 0, \ldots, l_{\max}$. Fig. 4.12 illustrates the delay-Doppler channel matrix with $M = 8$. The delay-Doppler channel block matrices $\mathbf{K}_{m,l}$ are related to $\tilde{\mathbf{K}}_{m,l}$ using 2D DFT operations,

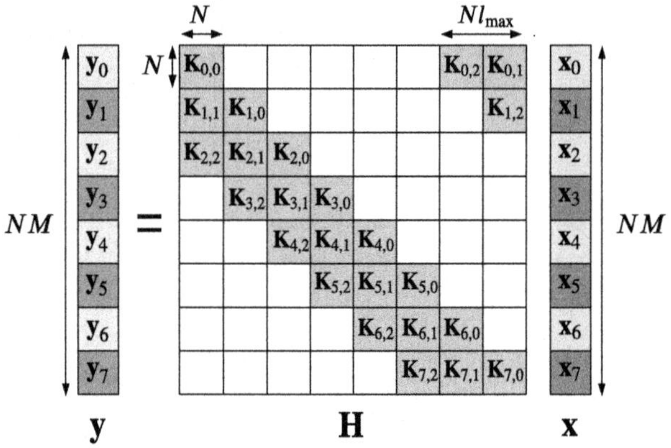

FIGURE 4.12 Delay-Doppler domain channel matrix $\mathbf{H} = (\mathbf{I}_M \otimes \mathbf{F}_N) \cdot \tilde{\mathbf{H}} \cdot (\mathbf{I}_M \otimes \mathbf{F}_N^{\dagger})$ with three delay paths partitioned into $N \times N$ submatrices.

$$\mathbf{K}_{m,l} = \mathbf{F}_N \cdot \tilde{\mathbf{K}}_{m,l} \cdot \mathbf{F}_N^{\dagger}. \tag{4.62}$$

Recalling the transmitted and received blocks (or subvectors) $\mathbf{x}_m,\ \mathbf{y}_m \in \mathbb{C}^{N\times 1}$ in (4.31), for $m = 0, \ldots, M-1$, we obtain the block-wise delay-Doppler input–output relation as

$$\boxed{\mathbf{y}_m = \sum_{l \in \mathcal{L}} \mathbf{K}_{m,l} \cdot \mathbf{x}_{[m-l]_M} + \mathbf{z}_m,} \tag{4.63}$$

where $\mathbf{z}_m \in \mathbb{C}^{N\times 1}$ is the AWGN vector.

Note that the noise vectors in all four domains $\mathbf{w}$ (time), $\check{\mathbf{w}}$ (time-frequency), $\tilde{\mathbf{w}}$ (delay-time), and $\mathbf{z}$ (delay-Doppler) have the same statistical properties, since they are related by unitary transformations.

4.5 Variants of OTFS

In the previous sections, we analyzed the OTFS transceiver and the input–output relation of a single OTFS time frame of NM samples in isolation. In this section, we discuss different variations of the OTFS frame structure (see Fig. 4.13).

- *Reduced ZP/CP-OTFS (RZP/RCP-OTFS)*: a single ZP is appended or a single CP is prepended to an OTFS frame (see Fig. 4.13(a))
- *ZP/CP-OTFS*: a ZP or CP of L_g samples is added (ZP appended and CP prepended) to each block of the time domain OTFS frame (see Fig. 4.13(b)).

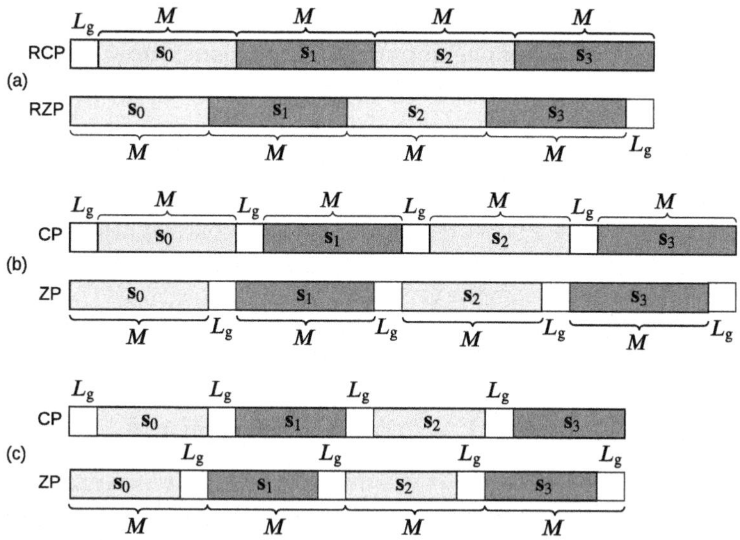

FIGURE 4.13 The time domain frame for different variants of OTFS (CPs are prepended and ZPs are appended). Option (a): a single CP/ZP is added to the frame; option (b): a CP/ZP is added to each block; option (c): a CP/ZP is included within each block.

In the case of Fig. 4.13(b), the OTFS frame duration T_f is extended by a factor $\gamma_g = (1 + L_g/M)$, and therefore the Doppler resolution $1/T_f$ increases. After discarding ZP/CP at the receiver, all the Doppler shifts κ_i of the channel taps get scaled by a factor of γ_g. For compatibility with the more commonly adopted cyclic prefix insertion in OFDM, we will follow this case in this chapter and in all discussions related to detection (Chapters 6 and 8).

An alternative solution is shown in Fig. 4.13(c), where the number of information symbols is reduced in the OTFS frame to $(M - L_g) \times N$, so that the total frame size remains MN samples, after inserting the ZP/CP. At the receiver, ZP/CP are not discarded, so that the Doppler shifts remain unscaled in the $M \times N$ delay-Doppler grid. We will follow this case in Chapter 7 when describing channel estimation methods and in most of the MATLAB codes given in Appendix C.

4.5.1 Reduced ZP OTFS

For RZP-OTFS, we consider an OTFS time frame of NM samples in isolation. We add a ZP of length l_{max} to the frame to avoid interference from the previous frame due to channel delay spread, as shown in Fig. 4.13. Note that RZP-OTFS differs from ZP-OTFS, since RZP-OTFS has one ZP per frame, while ZP-OTFS has one ZP per block.

RZP-OTFS: time domain analysis

The RZP-OTFS time domain input–output relation is the same as (4.36), i.e.,

$$\mathbf{r}[q]=\sum_{l} g^{\mathrm{s}}[l,q]\mathbf{s}[q-l]+\mathbf{w}[q], \quad q=0,\ldots,NM-1, \tag{4.64}$$

where $\mathbf{s}[q-l]=0$ for $q-l<0$ and $g^{\mathrm{s}}[l,q]$ is the channel response $g(\tau,t)$ sampled at $\tau=lT/M$ and $t=qT/M$. The vectorized input–output relation is the same as in (4.37), i.e., $\mathbf{r}=\mathbf{G}\cdot\mathbf{s}+\mathbf{w}$.

Next, we derive the element-wise input–output relation. Omitting noise and subsitituting $q=m+nM$ into (4.64) yields

$$\mathbf{r}[m+nM]=\sum_{l} g^{\mathrm{s}}[l,m+nM]\,\mathbf{s}[m+nM-l], \tag{4.65}$$

for $m=0,\ldots,M-1$ and $n=0,\ldots,N-1$.

We can now narrow down to each OTFS block. For the n-th block, the m-th Rx/Tx samples can be rewritten as

$$\mathbf{r}_n[m]=\mathbf{r}[m+nM], \tag{4.66}$$

and

$$\begin{aligned}\mathbf{s}_n[m-l]&=\mathbf{s}[m+nM-l] \text{ for } m\geq l,\\ \mathbf{s}_n\big[[m-l]_M\big]&=\mathbf{s}[m+nM-l] \text{ for } m<l.\end{aligned} \tag{4.67}$$

Then, the element-wise input–output relation can be obtained by substituting (4.66) and (4.67) into (4.65) yields

$$\mathbf{r}_n[m]=\sum_{l,l\leq m} g^{\mathrm{s}}[l,m+nM]\mathbf{s}_n[m-l]+\underbrace{\sum_{l,l>m} g^{\mathrm{s}}[l,m+nM]\mathbf{s}_{n-1}\big[[m-l]_M\big]}_{\text{interblock interference}}, \tag{4.68}$$

For the first block ($n=0$), we set the zero-padding

$$\mathbf{s}_{n-1}\big[[m-l]_M\big]=0 \quad \text{for } n=0 \text{ and } m<l,$$

which yields

$$\mathbf{r}_0[m]=\sum_{l,l\leq m} g^{\mathrm{s}}[l,m]\mathbf{s}_0[m-l], \tag{4.69}$$

implying that the first block $\mathbf{s}_0$ is free of interblock interference as no samples are transmitted before the 0-th block.

RZP-OTFS: delay-time domain analysis

Let us take a look at how the interblock interference manifests in the delay-time domain. Due to the row-column interleaving in (4.52), the delay-time block samples are related to the time domain block samples as

$$\mathbf{r}_n[m] = \tilde{\mathbf{y}}_m[n], \quad \mathbf{s}_n[m] = \tilde{\mathbf{x}}_m[n]. \tag{4.70}$$

In (4.68), replacing $g^{\mathrm{s}}[l, m + nM]$ with $\tilde{\boldsymbol{\nu}}_{m,l}[n]$ and replacing $\mathbf{r}_n[m]$, $\mathbf{s}_n$, and $\mathbf{s}_{n-1}$ using the relations in (4.70), we obtain the delay-time symbol vectors $\tilde{\mathbf{y}}_m[n]$ as

$$\tilde{\mathbf{y}}_m[n] = \sum_{l, l \le m} \tilde{\boldsymbol{\nu}}_{m,l}[n] \tilde{\mathbf{x}}_{m-l}[n] + \sum_{l, l > m} \tilde{\boldsymbol{\nu}}_{m,l}[n] \tilde{\mathbf{x}}_{[m-l]_M}[n-1], \tag{4.71}$$

for $m = 0, \ldots, M-1$ and $n = 0, \ldots, N-1$, where we set

$$\tilde{\mathbf{x}}_{[m-l]_M}[n-1] = 0 \quad \text{for } n = 0 \text{ and } m < l$$

and the delay-time channel $\tilde{\boldsymbol{\nu}}_{m,l} \in \mathbb{C}^{N \times 1}$ is expressed based on (4.6) as

$$\begin{aligned}
\tilde{\boldsymbol{\nu}}_{m,l}[n] &= g^{\mathrm{s}}[l, q] = g^{\mathrm{s}}[l, m + nM] \\
&= \sum_{\ell \in \mathcal{L}} \left(\sum_{\kappa \in \mathcal{K}_\ell} \nu_\ell(\kappa) z^{\kappa(m+nM-l)} \right) \mathrm{sinc}(l - \ell) \\
&= \sum_{\ell \in \mathcal{L}} \left(\sum_{\kappa \in \mathcal{K}_\ell} \nu_\ell(\kappa) z^{\kappa(m-l)} e^{\frac{j2\pi\kappa n}{N}} \right) \mathrm{sinc}(l - \ell).
\end{aligned} \tag{4.72}$$

In the case of integer delay tap channels, i.e., $\ell = l \in \mathbb{Z}$, (4.72) reduces to

$$\tilde{\boldsymbol{\nu}}_{m,l}[n] = \sum_{\kappa \in \mathcal{K}_l} \nu_l(\kappa) z^{\kappa(m-l)} e^{\frac{j2\pi\kappa n}{N}} \quad n = 0, \ldots, N-1, \tag{4.73}$$

which shows that $\tilde{\boldsymbol{\nu}}_{m,l}$ for each delay tap l is the inverse Fourier transform of $\nu_l(\kappa) z^{\kappa(m-l)}$. The delay-time input–output relation in (4.71) including noise can be generalized as

$$\tilde{\mathbf{y}}_m = \sum_{l \in \mathcal{L}} \tilde{\mathbf{K}}_{m,l} \cdot \tilde{\mathbf{x}}_{[m-l]_M} + \tilde{\mathbf{w}}_m \quad m = 0, \ldots M-1, \tag{4.74}$$

where $\tilde{\mathbf{w}}_m \in \mathbb{C}^{N \times 1}$ is the AWGN vector and $\tilde{\mathbf{K}}_{m,l} \in \mathbb{C}^{N \times N}$ are the submatrices of $\tilde{\mathbf{H}}$, representing the delay-time channel matrix between the $([m-l]_M)$-th transmitted and m-th received delay-time vector, given by

$$\tilde{\mathbf{K}}_{m,l} = \begin{cases} \mathrm{diag}[\tilde{\boldsymbol{\nu}}_{m,l}], & \text{if } m \ge l, \\ \mathrm{diag}[0, \tilde{\boldsymbol{\nu}}_{m,l}[1], \cdots, \tilde{\boldsymbol{\nu}}_{m,l}[N-1]] \cdot \boldsymbol{\Pi}, & \text{if } m < l, \end{cases} \tag{4.75}$$

where $\boldsymbol{\Pi}$ is the permutation matrix for column-wise left cyclic shift. The matrix $\tilde{\mathbf{K}}_{m,l}$ is formed as

$$\underbrace{\begin{bmatrix} \tilde{\boldsymbol{\nu}}_{m,l}[0] & 0 & \cdots & 0 \\ 0 & \tilde{\boldsymbol{\nu}}_{m,l}[1] & \cdots & 0 \\ \vdots & \ddots & \ddots & \vdots \\ 0 & 0 & \cdots & \tilde{\boldsymbol{\nu}}_{m,l}[N-1] \end{bmatrix}}_{\tilde{\mathbf{K}}_{m,l} \text{ if } m \geq l} \underbrace{\begin{bmatrix} 0 & 0 & \cdots & 0 \\ \tilde{\boldsymbol{\nu}}_{m,l}[1] & 0 & \cdots & 0 \\ \vdots & \ddots & \ddots & \vdots \\ 0 & \cdots & \tilde{\boldsymbol{\nu}}_{m,l}[N-1] & 0 \end{bmatrix}}_{\tilde{\mathbf{K}}_{m,l} \text{ if } m<l}.$$

Fig. 4.14 shows the delay-time channel matrix for RZP-OTFS with $M = 8$. The case when $m < l$ corresponds to the top-right corner blocks (shaded blocks in Fig. 4.14) of the delay-time channel matrix $\tilde{\mathbf{H}}$.

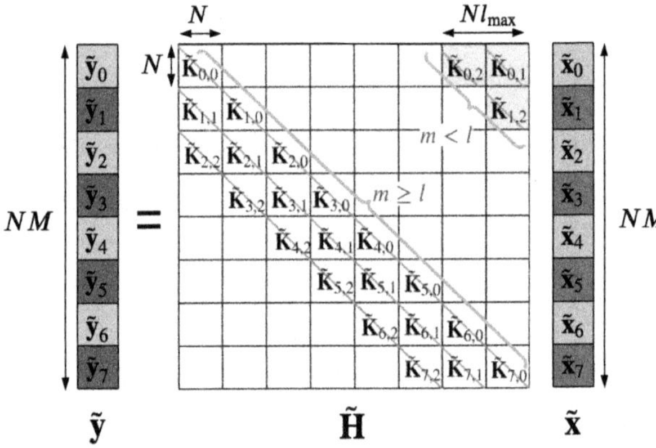

FIGURE 4.14 RZP-OTFS delay-time domain channel matrix $\tilde{\mathbf{H}} = \mathbf{P}^{\mathrm{T}} \cdot \mathbf{G} \cdot \mathbf{P}$ with three delay paths partitioned into $N \times N$ submatrices.

RZP-OTFS: delay-Doppler domain analysis

Using the relation between the delay-time and delay-Doppler given in (4.62), the block-wise input–output relation in (4.74) can be converted to delay-Doppler domain as

$$\mathbf{y}_m = \sum_{l \in \mathcal{L}} \mathbf{K}_{m,l} \cdot \mathbf{x}_{[m-l]_M} + \mathbf{z}_m \quad m = 0, \ldots M-1,$$

where $\mathbf{z}_m \in \mathbb{C}^{N \times 1}$ is the delay-Doppler AWGN vector and

$$\mathbf{K}_{m,l} = \begin{cases} \mathrm{circ}[\boldsymbol{\nu}_{m,l}], & \text{if } m \geq l, \\ (\mathrm{circ}[\boldsymbol{\nu}_{m,l}] - \frac{1}{N}\tilde{\boldsymbol{\nu}}_{m,l}(0)\mathbf{1}_{N\times N}) \cdot \mathbf{D}, & \text{if } m < l \end{cases} \quad (4.76)$$

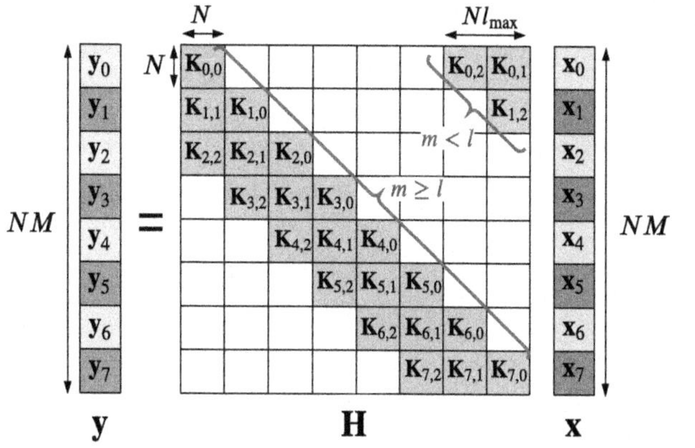

FIGURE 4.15 RZP-OTFS delay-Doppler domain channel matrix $\mathbf{H}$ with three delay paths partitioned into $N \times N$ submatrices.

is the Doppler spread matrix at the l-th delay tap, where $\mathbf{1}_{N\times N} \in \mathbb{C}^{N\times N}$ denotes the all-ones matrix and

$$\mathbf{D} = \mathrm{diag}\left[1, e^{-j2\pi/N}, \cdots, e^{-j2\pi(N-1)/N}\right]$$

is the phase rotation[2] due to the cyclic shift of $\tilde{\mathbf{K}}_{m,l}$ with the $\boldsymbol{\Pi}$ matrix in (4.75). The circulant matrix $\mathrm{circ}[\boldsymbol{\nu}_{m,l}]$ is formed as

$$\mathrm{circ}[\boldsymbol{\nu}_{m,l}] = \begin{bmatrix} \boldsymbol{\nu}_{m,l}[0] & \boldsymbol{\nu}_{m,l}[N-1] & \cdots & \boldsymbol{\nu}_{m,l}[1] \\ \boldsymbol{\nu}_{m,l}[1] & \boldsymbol{\nu}_{m,l}[0] & \cdots & \boldsymbol{\nu}_{m,l}[2] \\ \vdots & \ddots & \ddots & \vdots \\ \boldsymbol{\nu}_{m,l}[N-1] & \boldsymbol{\nu}_{m,l}[N-2] & \cdots & \boldsymbol{\nu}_{m,l}[0] \end{bmatrix}.$$

The delay-time channel vectors $\tilde{\boldsymbol{\nu}}_{m,l}$ are related to the delay-Doppler channel vectors $\boldsymbol{\nu}_{m,l}$ using the 1D Fourier transformation as

$$\boldsymbol{\nu}_{m,l}[k] = \frac{1}{N}\sum_{n=0}^{N-1} \tilde{\boldsymbol{\nu}}_{m,l}[n] e^{\frac{-j2\pi kn}{N}}. \tag{4.77}$$

Fig. 4.15 shows the delay-Doppler domain channel matrix for RZP-OTFS with $M = 8$. Note that for $m \geq l$, $\mathbf{K}_{m,l}$ is a circulant matrix, and for $m < l$ (top-right corner blocks), it is not.

[2]Cyclic shift of a 2D signal in the time domain leads to a phase shift in the corresponding 2D Fourier domain, i.e., for any square matrix $\mathbf{A} \in \mathbb{C}^{N\times N}$: $\mathbf{F}_N \cdot (\mathbf{A} \cdot \boldsymbol{\Pi}^i) \cdot \mathbf{F}_N^{\dagger} = (\mathbf{F}_N \cdot \mathbf{A} \cdot \mathbf{F}_N^{\dagger}) \cdot \mathbf{D}^i$. See the Appendix for the definition of the permutation matrix $\boldsymbol{\Pi}$.

Substituting (4.97) in (4.77), we now can write the discrete Doppler spread vector $\boldsymbol{\nu}_{m,l} \in \mathbb{C}^{N\times 1}$ in terms of the channel Doppler response $\nu_\ell(\kappa)$, for the following channel conditions.

RZP-OTFS: fractional delay and fractional Doppler shifts

Considering the case of the fractional delay and fractional Doppler shifts, we have

$$\boldsymbol{\nu}_{m,l}[k] = \frac{1}{N}\sum_{\ell\in\mathcal{L}}\Big(\sum_{\kappa\in\mathcal{K}_\ell} \nu_\ell(\kappa) z^{\kappa(m-l)} \zeta_N(\kappa - k)\Big)\operatorname{sinc}(l-\ell), \tag{4.78}$$

where $\ell, \kappa \in \mathbb{R}$ and the periodic sinc function $\zeta(\cdot)$ includes the extra phase and magnitude variations in the Doppler spread vectors due to fractional Doppler shifts, given as

$$\zeta_N(x) = \sum_{n=0}^{N-1} \mathrm{e}^{\frac{j2\pi x n}{N}} = \frac{\sin(\pi x)}{\sin(\pi x/N)}\mathrm{e}^{\frac{j\pi x(N-1)}{N}}. \tag{4.79}$$

RZP-OTFS: integer delay and fractional Doppler shifts

For integer values of $(l-\ell)$, the function $\operatorname{sinc}(l-\ell)$ evaluates to 1 when $\ell = l$ and zero elsewhere. Hence, (4.78) reduces to

$$\boldsymbol{\nu}_{m,l}[k] = \frac{1}{N}\sum_{\kappa\in\mathcal{K}_l} \nu_l(\kappa) z^{\kappa(m-l)} \zeta_N(\kappa - k) \tag{4.80}$$

for $l = \ell \in \mathbb{Z}$ and $\kappa \in \mathbb{R}$.

RZP-OTFS: integer delay and integer Doppler shifts

For integer values of x, the function $\zeta_N(x)$ evaluates to N when $x = 0$ and zero elsewhere. Hence (4.80) reduces to the simple form

$$\boldsymbol{\nu}_{m,l}[k] = \begin{cases} \nu_l(\kappa) z^{\kappa(m-l)}, & \text{if } l = \ell \text{ and } k = [\kappa]_N, \\ 0, & \text{otherwise,} \end{cases} \tag{4.81}$$

for $l = \ell \in \mathbb{Z}$ and $\kappa \in \mathbb{Z}$. Since κ can be a negative integer Doppler shift, we ensure $k = [\kappa]_N \in [0, N-1]$.

Fig. 4.16 shows the discrete Doppler spread vector $\boldsymbol{\nu}_{m,l}$ in the case for integer Doppler and fractional Doppler cases. The black dashed lines denote the periodic sinc function $\zeta_N(\cdot)$ defined in (4.79). It can be seen that in the case of fractional Doppler shifts, discretization of the Doppler indices causes leakage into all Doppler bins.

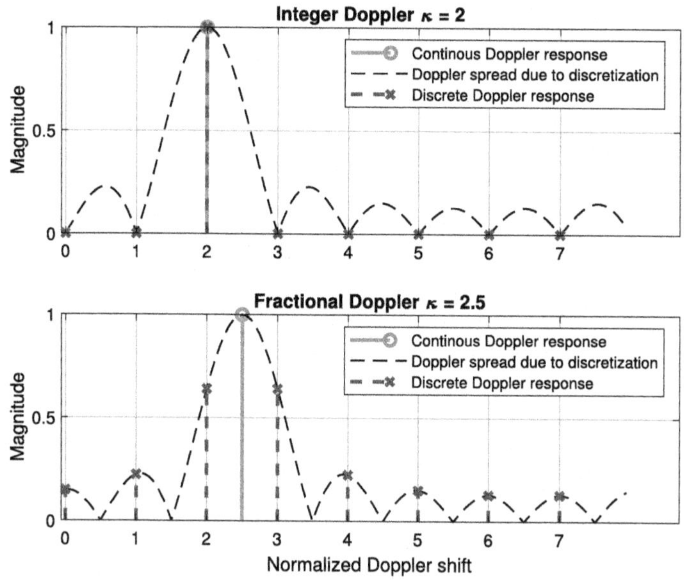

FIGURE 4.16 Continuous ($\nu_\ell(\kappa)$) vs. discrete ($\nu_{m,l}(k)$) Doppler response for integer Doppler $\mathcal{K}_\ell = \{2\}$ (top) and fractional Doppler $\mathcal{K}_\ell = \{2.5\}$ (bottom).

4.5.2 Reduced CP-OTFS

For RCP-OTFS, we prepend a CP of length $l_{\max}$ to the time domain OTFS frame, where the CP is the copy of the last $l_{\max}$ samples in the frame. At the receiver, the CP is discarded prior to further processing.

RCP-OTFS: time domain analysis

For RCP-OTFS, after CP removal, the time domain input–output relation in (4.36) can be modified as

$$\mathbf{r}[q] = \sum_{l} g^{\mathrm{s}}[l,q]\mathbf{s}\big[[q-l]_{MN}\big] + \mathbf{w}[q] \quad q = 0,\ldots,NM-1. \tag{4.82}$$

Note that the main difference in (4.82) with respect to the input–output relation in (4.36) is the modulo-MN operation due to CP removal. The vectorized RCP-OTFS input–output relation can be written as in (4.37), $\mathbf{r} = \mathbf{G}\cdot\mathbf{s} + \mathbf{w}$, where

$$\mathbf{G}\big[q,[q-l]_{MN}\big] = g^{\mathrm{s}}[l,q] \tag{4.83}$$

for $q = 0,\ldots,NM-1$.

For RCP-OTFS, MATLAB implementation on generating $\mathbf{G}$ using (4.83) and the received time domain signal is given in Appendix C (see MATLAB code 16).

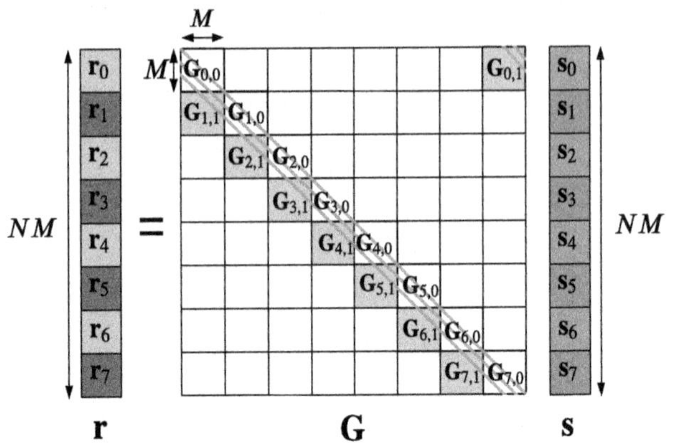

FIGURE 4.17 RCP-OTFS time domain channel matrix **G** with three delay paths partitioned into $M \times M$ submatrices.

Fig. 4.17 shows the time domain channel matrix for RCP-OTFS with $N = 8$. Unlike the case in Fig. 4.8, **G** is not a lower triangular matrix and has nonzero elements in the top-right corner due to CP. The block-wise input–output relation in (4.39) is reduced to

$$\mathbf{r}_n = \mathbf{G}_{n,0} \cdot \mathbf{s}_n + \mathbf{G}_{n,1} \cdot \mathbf{s}_{[n-1]_N} + \mathbf{w}_n. \tag{4.84}$$

Alternatively, the block-wise input–output relation (omitting noise for brevity) in (4.84) can be written in the element-wise format as

$$\mathbf{r}_n[m] = \sum_{l, l \le m} g^{\mathrm{s}}[l, m+nM]\mathbf{s}_n[m-l] + \underbrace{\sum_{l, l > m} g^{\mathrm{s}}[l, m+nM]\mathbf{s}_{[n-1]_N}\big[[m-l]_M\big]}_{\text{interblock interference}} \tag{4.85}$$

for $n = 0, \ldots, N-1$.

Unlike (4.68) for RZP-OTFS, the first block $\mathbf{s}_0$ will have interference from the CP, which is equivalent to having interference from the last block $\mathbf{s}_{N-1}$.

RCP-OTFS: delay-time and delay-Doppler domain analysis

Following the same steps in the derivation of RZP-OTFS, we can rewrite (4.85) in vectorized form in the delay-time domain as

$$\tilde{\mathbf{y}}_m = \sum_{l \in \mathcal{L}} \tilde{\mathbf{K}}_{m,l} \cdot \tilde{\mathbf{x}}_{[m-l]_M} + \tilde{\mathbf{w}}_m, \tag{4.86}$$

where

$$\tilde{\mathbf{K}}_{m,l} = \begin{cases} \text{diag}[\tilde{\boldsymbol{\nu}}_{m,l}], & \text{if } m \geq l, \\ \text{diag}[\tilde{\boldsymbol{\nu}}_{m,l}] \cdot \boldsymbol{\Pi}, & \text{if } m < l \end{cases} \tag{4.87}$$

is the delay-time channel matrix between the $([m-l]_M)$-th transmitted and m-th received delay-time vector. The matrices $\tilde{\mathbf{K}}_{m,l}$ for RCP-OTFS are formed as

$$\underbrace{\begin{bmatrix} \tilde{\nu}_{m,l}(0) & 0 & \cdots & 0 \\ 0 & \tilde{\nu}_{m,l}(1) & \cdots & 0 \\ \vdots & \ddots & \ddots & \vdots \\ 0 & 0 & \cdots & \tilde{\nu}_{m,l}(N-1) \end{bmatrix}}_{\tilde{\mathbf{K}}_{m,l} \text{ if } m \geq l} \underbrace{\begin{bmatrix} 0 & 0 & \cdots & \tilde{\nu}_{m,l}(0) \\ \tilde{\nu}_{m,l}(1) & 0 & \cdots & 0 \\ \vdots & \ddots & \ddots & \vdots \\ 0 & \cdots & \tilde{\nu}_{m,l}(N-1) & 0 \end{bmatrix}}_{\tilde{\mathbf{K}}_{m,l} \text{ if } m<l}.$$

Similarly, (4.87) can be converted to the delay-Doppler domain as

$$\mathbf{y}_m = \sum_{l \in \mathcal{L}} \mathbf{K}_{m,l} \cdot \mathbf{x}_{[m-l]_M} + \mathbf{z}_m, \tag{4.88}$$

where the Doppler spread matrix at the l-th delay tap

$$\mathbf{K}_{m,l} = \begin{cases} \text{circ}[\boldsymbol{\nu}_{m,l}], & \text{if } m \geq l, \\ \text{circ}[\boldsymbol{\nu}_{m,l}] \cdot \mathbf{D}, & \text{if } m < l. \end{cases} \tag{4.89}$$

Adding a CP instead of ZP simplifies the input–output relation compared to (4.76). Note that, similar to RZP-OTFS, for $m \geq l$, $\mathbf{K}_{m,l}$ is a circulant matrix, while for $m < l$, it is not. It is convenient to have all the submatrices $\mathbf{K}_{m,l}$ of the delay-Doppler channel matrix $\mathbf{H}$ as circulant blocks, since this allows the use of low complexity detection methods. To obtain such effect, we discuss the next two OTFS transmission schemes.

4.5.3 CP-OTFS

In this section, we present the CP-OTFS scheme, where a CP is prepended to each of the N transmitted time domain blocks, similar to CP-OFDM. At the receiver, the N CPs are discarded prior to further processing.

CP-OTFS: time domain analysis

Let $\mathbf{s}_{\text{CP}}$ and $\mathbf{r}_{\text{CP}}$ be the time domain transmitted and received signals after adding the CP. Then the time domain input–output relation can be written as

$$\mathbf{r}_{\text{CP}}[q] = \sum_{l \in \mathcal{L}} g^{\text{s}}[l,q] \mathbf{s}_{\text{CP}}[q-l] + \mathbf{w}_{\text{CP}}[q], \tag{4.90}$$

where $q = 0, \ldots, (M + L_{\text{CP}})N - 1$ and $L_{\text{CP}} \geq l_{\max}$ denotes the length of the CP.

After CP removal in each block, (4.90) becomes

$$\mathbf{r}[m + nM] = \sum_{l \in \mathcal{L}} g^{\text{s}}\big[l, m + n(M + L_{\text{CP}})\big]\mathbf{s}\big[[m - l]_M + nM\big] + \mathbf{w}[m + nM] \tag{4.91}$$

for $m = 0, \ldots, M - 1$ and $n = 0, \ldots, N - 1$, where the indices in g^{s} and the modulo-M operation are due to CP removal per block.

The block-wise matrix input–output relation can be written as

$$\mathbf{r}_n = \mathbf{G}_{n,0} \cdot \mathbf{s}_n + \mathbf{w}_n, \tag{4.92}$$

where $\mathbf{r}_n, \mathbf{s}_n, \mathbf{w}_n \in \mathbb{C}^{M \times 1}$ are the subvectors of the received and transmitted samples, as well as the AWGN noise, and $\mathbf{G}_{n,0} \in \mathbb{C}^{M \times M}$, $n = 0, \ldots, N - 1$, are the diagonal blocks of channel matrix $\mathbf{G}$ (see an example illustration in Fig. 4.18 with N), with elements

$$\mathbf{G}_{n,0}\big[m, [m - l]_M\big] = g^{\text{s}}\big[l, m + n(M + L_{\text{CP}})\big]. \tag{4.93}$$

As shown in Fig. 4.18 with $N = 8$, $\mathbf{G}_{n,0}$ is a banded matrix with a bandwidth of $l_{\max} + 1$ due to CP per block. Unlike the RZP/RCP-OTFS in (4.68) and (4.85), there is *no interblock interference* for CP-OTFS thanks to the CP per block.

The code for generating the equivalent $\mathbf{G}$ matrix together with the received time domain CP-OTFS signal after CP removal per block is given in MATLAB code 17 of Appendix C.

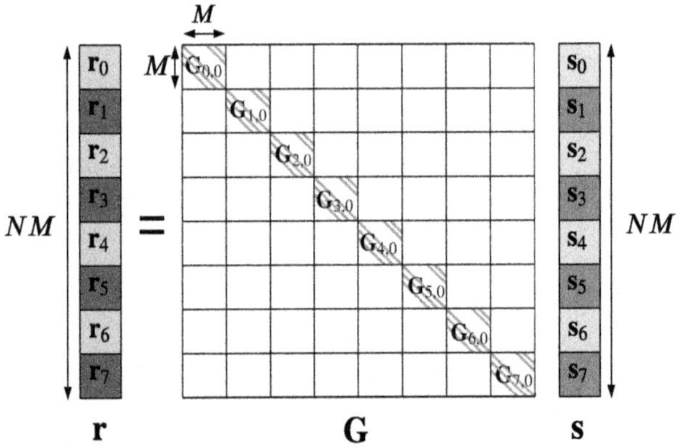

FIGURE 4.18 CP-OTFS time domain channel matrix $\mathbf{G}$ with three delay paths partitioned into $M \times M$ submatrices.

CP-OTFS: delay-time domain analysis

Now consider each block of CP-OTFS. Based on (4.93), the input–output relation in (4.92) (omitting noise for simplification) can be expressed in the element-wise format as

$$\begin{aligned} \mathbf{r}_n[m] &= \sum_{l\in\mathcal{L}} \mathbf{G}_{n,0}\big[m, [m-l]_M\big]\mathbf{s}_n\big[[m-l]_M\big] \\ &= \sum_{l\in\mathcal{L}} g^{\mathrm{s}}\big[l, m+n(M+L_{\mathrm{CP}})\big]\mathbf{s}_n\big[[m-l]_M\big] \end{aligned} \tag{4.94}$$

for $n = 0, \ldots, N-1$ and $m = 0, \ldots, M-1$. Using the row-column interleaving in (4.52), the delay-time block samples are related to the time domain block samples as

$$\mathbf{r}_n[m] = \tilde{\mathbf{y}}_m[n], \quad \mathbf{s}_n[m] = \tilde{\mathbf{x}}_m[n]. \tag{4.95}$$

Similar to the derivation for RCP-OTFS, by replacing $g^{\mathrm{s}}[l, m+n(M+L_{\mathrm{CP}})]$ with $\tilde{\boldsymbol{\nu}}_{m,l}[n]$ and replacing elements of $\mathbf{s}_n$ using the relations in (4.95), the delay-time version of (4.94) is given by

$$\tilde{\mathbf{y}}_m[n] = \sum_{l\in\mathcal{L}} \tilde{\boldsymbol{\nu}}_{m,l}[n]\tilde{\mathbf{x}}_{[m-l]_M}[n] \tag{4.96}$$

for $n = 0, \ldots, N-1$ and $m = 0, \ldots, M-1$, where $\tilde{\boldsymbol{\nu}}_{m,l} \in \mathbb{C}^{N\times 1}$ is

$$\tilde{\boldsymbol{\nu}}_{m,l}[n] = \sum_{\ell\in\mathcal{L}}\Bigg(\sum_{\kappa\in\mathcal{K}_l} \nu_\ell(\kappa) z^{\kappa(m-l)} e^{\frac{j2\pi}{N}\kappa\gamma_{\mathrm{g}} n}\Bigg)\mathrm{sinc}(l-\ell), \quad n = 0, \ldots, N-1, \tag{4.97}$$

where $\gamma_{\mathrm{g}} = (1+\frac{L_{\mathrm{CP}}}{M})$ is the scaling of the actual channel Doppler shifts due to the increase in the duration of the frame as a result of CP between the blocks. Assuming integer delay tap channels, i.e., $l = \ell \in \mathbb{Z}$, (4.97) reduces to

$$\tilde{\boldsymbol{\nu}}_{m,l}[n] = \sum_{\kappa\in\mathcal{K}_l} \nu_l(\kappa) z^{\kappa(m-l)} e^{\frac{j2\pi}{N}\kappa\gamma_{\mathrm{g}} n} \quad n = 0, \ldots, N-1. \tag{4.98}$$

We can note from (4.98) that the discrete delay-time response $\tilde{\boldsymbol{\nu}}_{m,l}$ for each delay tap l at time instants $t = m\frac{T}{M} + nT\gamma_{\mathrm{g}}$ is related to the inverse Fourier transform of $\nu_l(\kappa)z^{\kappa(q-l)}$ sampled at $q = m + nM\gamma_{\mathrm{g}}$.

Let $\tilde{\mathbf{K}}_{m,l} \in \mathbb{C}^{N\times N}$ be the delay-time channel matrix between the $[m-l]_M$-th transmitted and m-th received delay-time symbol vector and

$$\tilde{\mathbf{K}}_{m,l} = \mathrm{diag}[\tilde{\boldsymbol{\nu}}_{m,l}], \quad m = 0, \ldots, M-1. \tag{4.99}$$

Using (4.99), the delay-time input–output relation in (4.96) can be written as

$$\tilde{\mathbf{y}}_m = \sum_{l\in\mathcal{L}} \tilde{\mathbf{K}}_{m,l} \cdot \tilde{\mathbf{x}}_{[m-l]_M} = \sum_{l\in\mathcal{L}} \mathrm{diag}[\tilde{\boldsymbol{\nu}}_{m,l}] \cdot \tilde{\mathbf{x}}_{[m-l]_M}$$
$$= \sum_{l\in\mathcal{L}} (\tilde{\boldsymbol{\nu}}_{m,l} \circ \tilde{\mathbf{x}}_{[m-l]_M}), \tag{4.100}$$

where the operation $\mathbf{a} \circ \mathbf{b}$ denotes element-wise multiplication vectors $\mathbf{a}$ and $\mathbf{b}$.

CP-OTFS: delay-Doppler domain analysis

Using (4.100), the delay-Doppler domain received symbol vectors can be obtained as

$$\mathbf{y}_m = \mathbf{F}_N \cdot \tilde{\mathbf{y}}_m = \sum_{l\in\mathcal{L}} \mathbf{F}_N \cdot (\tilde{\boldsymbol{\nu}}_{m,l} \circ \tilde{\mathbf{x}}_{[m-l]_M})$$
$$= \sum_{l\in\mathcal{L}} (\mathbf{F}_N \cdot \tilde{\boldsymbol{\nu}}_{m,l}) \circledast (\mathbf{F}_N \cdot \tilde{\mathbf{x}}_{[m-l]_M})$$
$$= \sum_{l\in\mathcal{L}} \boldsymbol{\nu}_{m,l} \circledast \mathbf{x}_{[m-l]_M}, \tag{4.101}$$

where operator $\circledast$ denotes circular convolution, the second step is due to the circular convolution property of Fourier transform in Appendix B, and

$$\boldsymbol{\nu}_{m,l}[k] = \frac{1}{N} \sum_{n=0}^{N-1} \tilde{\boldsymbol{\nu}}_{m,l}(n) e^{\frac{-j2\pi kn}{N}} \tag{4.102}$$

for $0 \le k \le N-1$, $0 \le m \le M-1$, and $\boldsymbol{\nu}_{m,l}$ is the discrete Doppler spread vector in the l-th channel delay tap, experienced by all the symbols in the $([m-l]_M)$-th row of the $M \times N$ delay-Doppler grid.

Then (4.101) can be written as

$$\mathbf{y}_m = \sum_{l\in\mathcal{L}} \mathbf{K}_{m,l} \cdot \mathbf{x}_{[m-l]_M}, \tag{4.103}$$

where the Doppler spread matrix at the l-th delay tap is

$$\mathbf{K}_{m,l} = \mathrm{circ}[\boldsymbol{\nu}_{m,l}], \quad m = 0, \ldots, M-1. \tag{4.104}$$

Substituting (4.97) in (4.102), we can write the discrete Doppler spread vector $\boldsymbol{\nu}_{m,l} \in \mathbb{C}^{N\times 1}$ in terms of the channel Doppler response $\nu_\ell(\kappa)$, for the following channel models.

CP-OTFS: fractional delay and fractional Doppler shifts

In the case of fractional delay and fractional Doppler shifts, we obtain

$$\boldsymbol{\nu}_{m,l}[k] = \frac{1}{N}\sum_{\ell\in\mathcal{L}}\Big(\sum_{\kappa\in\mathcal{K}_\ell} \nu_\ell(\kappa) z^{\kappa(m-l)} \zeta_N(\kappa\gamma_g - k)\Big)\mathrm{sinc}(l-\ell), \tag{4.105}$$

where $\ell, \kappa \in \mathbb{R}$ and $\gamma_g = (1 + \frac{L_{CP}}{M})$.

CP-OTFS: integer delay and fractional Doppler shifts

For integer values of $(l - \ell)$, the function $\mathrm{sinc}(l - \ell)$ evaluates to 1 when $\ell = l$ and zero elsewhere. Hence, (4.105) reduces to

$$\boldsymbol{\nu}_{m,l}[k] = \frac{1}{N}\sum_{\kappa\in\mathcal{K}_l} \nu_l(\kappa) z^{\kappa(m-l)} \zeta_N(\kappa\gamma_g - k) \tag{4.106}$$

for $\ell = l \in \mathbb{Z}$ and $\kappa \in \mathbb{R}$.

CP-OTFS: integer delay and integer Doppler shifts

For integer values of x, the function $\zeta_N(x)$ evaluates to $\sqrt{N}$ when $x = 0$ and zero elsewhere. Hence (4.106) reduces to

$$\boldsymbol{\nu}_{m,l}[k] = \begin{cases} \nu_l(\kappa) z^{\kappa(m-l)}, & \text{if } \ell = l \text{ and } k = [\kappa\gamma_g]_N, \\ 0, & \text{otherwise} \end{cases} \tag{4.107}$$

for $\ell = l \in \mathbb{Z}$ and $\kappa\gamma_g \in \mathbb{Z}$. Since $\kappa\gamma_g$ can be a negative integer Doppler shift, we ensure $k = [\kappa\gamma_g]_N \in [0, N-1]$.

4.5.4 ZP-OTFS

In this section, we introduce ZP-OTFS, where a ZP of length $L_{ZP} \geq l_{max}$ is added to each of the N time domain blocks. One advantage of ZP-OTFS is that the ZP enables inserting pilot symbols for channel estimation, which will be presented in Chapter 7.

ZP-OTFS: time domain analysis

Similar to (4.92), the block-wise time domain input–output relation of a noiseless ZP-OTFS after removing the ZP is given as

$$\mathbf{r}_n = \mathbf{G}_{n,0} \cdot \mathbf{s}_n, \tag{4.108}$$

where $\mathbf{G}_{n,0} \in \mathbb{C}^{M\times M}$, $n = 0, \ldots, N-1$, are the diagonal blocks of channel matrix $\mathbf{G}$ (see an example illustration in Fig. 4.19 with $N = 8$) with elements

$$\mathbf{G}_{n,0}[m, m-l] = g^{\mathrm{s}}\big[l, m + n(M + L_{ZP})\big] \quad \text{for } m \geq l. \tag{4.109}$$

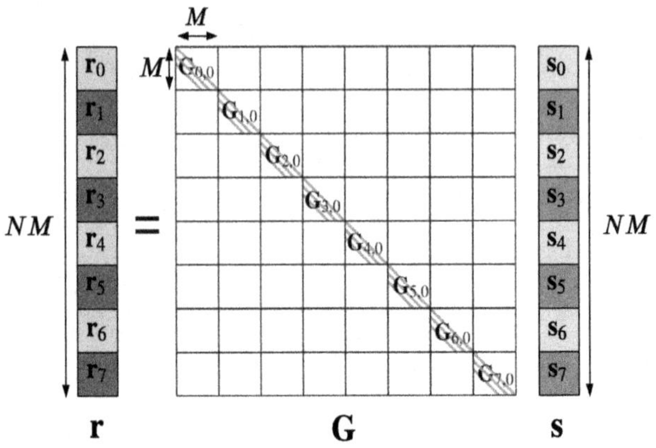

FIGURE 4.19 ZP-OTFS time domain channel matrix $\mathbf{G}$ with three delay paths partitioned into $M \times M$ submatrices.

for $l \in \mathcal{L}$ and *zero* otherwise, i.e., $\mathbf{G}_{n,0}$ are banded lower triangular matrices with a bandwidth $(l_{\max} + 1)$ due to the ZP per block.

The code for generating the $\mathbf{G}$ matrix and received time domain ZP-OTFS signal after ZP removal per block is given in MATLAB code 18 of Appendix C.

ZP-OTFS: delay-time and delay-Doppler domain analysis

The delay-time domain input–output relation is then modified as

$$\tilde{\mathbf{y}}_m[n] = \sum_{l \in \mathcal{L}} \tilde{\boldsymbol{\nu}}_{m,l}[n] \tilde{\mathbf{x}}_{m-l}[n], \tag{4.110}$$

where $m = 0, \ldots, M - 1$ and

$$\tilde{\mathbf{x}}_{m-l}[n] = 0 \quad \text{for } m < l. \tag{4.111}$$

Similar to CP-OTFS, there is *no interblock interference* for ZP-OTFS thanks to the ZP per block.

Following the same steps as in (4.101), the delay-Doppler input–output relation is then modified as

$$\mathbf{y}_m = \sum_{l \in \mathcal{L}} \mathbf{K}_{m,l} \cdot \mathbf{x}_{m-l} = \sum_{l \in \mathcal{L}} \boldsymbol{\nu}_{m,l} \circledast \mathbf{x}_{m-l}, \tag{4.112}$$

where $m = 0, \ldots, M - 1$ and

$$\mathbf{x}_{m-l}[n] = 0 \quad \text{for } m < l. \tag{4.113}$$

Similar to CP-OTFS, we can rewrite the discrete Doppler spread vector $\boldsymbol{\nu}_{m,l} \in \mathbb{C}^{N\times 1}$ in terms of the channel Doppler response $\nu_{\ell}(\kappa)$ for various channels by replacing $L_{\rm CP}$ by $L_{\rm ZP}$ in (4.105), (4.106), and (4.107).

4.6 Summary of channel representations and input–output relations for OTFS variants

In this chapter, we introduced four OTFS variants following the order of the *OTFS technology timeline*. We first introduced RZP/RCP-OTFS and then CP/ZP-OTFS, since the former were the original OTFS schemes proposed by Hadani et al.

In this section, we provide a summary of channel representations and input–output relations for these schemes. For the purpose of easy explanation, we present our summary in the order of *increasing complexity in mathematical expressions*. We start with the simplest scheme, CP-OTFS, followed by ZP-OTFS, RCP-OTFS, and RZP-OTFS. MATLAB codes for generating the channel matrix $\mathbf{G}$ and the received time domain signal $\mathbf{r}$ for all cases are given in Appendix C.

4.6.1 Channel representations for OTFS variants

From previous sections, the delay-time channel vector $\tilde{\boldsymbol{\nu}}_{m,l}(n)$ can be obtained from the discrete time domain channel $g^{\rm s}[q,l]$ as

$$\tilde{\boldsymbol{\nu}}_{m,l}[n] = \begin{cases} g^{\rm s}[m+n(M+L_{\rm CP}),l] & \text{CP-OTFS,} \\ g^{\rm s}[m+n(M+L_{\rm ZP}),l] & \text{ZP-OTFS,} \\ g^{\rm s}[m+nM,l] & \text{RCP-OTFS and RZP-OTFS} \end{cases}$$

for $n=0,\ldots,N-1$, $m=0,\ldots,M-1$, and $l \in \mathcal{L}$. Then, the delay-Doppler channel vectors can be obtained as

$$\boldsymbol{\nu}_{m,l}[k] = \frac{1}{N}\sum_{n=0}^{N-1} \tilde{\boldsymbol{\nu}}_{m,l}[n] e^{\frac{-j2\pi kn}{N}}.$$

The block-wise delay-time and delay-Doppler input–output relations are generalized for all cases as

$$\boxed{\tilde{\mathbf{y}}_m = \sum_{l\in\mathcal{L}} \tilde{\mathbf{K}}_{m,l} \cdot \tilde{\mathbf{x}}_{[m-l]_M} + \tilde{\mathbf{w}}_m,} \tag{4.114}$$

$$\boxed{\mathbf{y}_m = \sum_{l\in\mathcal{L}} \mathbf{K}_{m,l} \cdot \mathbf{x}_{[m-l]_M} + \mathbf{z}_m.} \tag{4.115}$$

The above delay-Doppler domain input–output relation is illustrated in Fig. 4.20 using a factor graph for an example with three delay paths, $l \in$

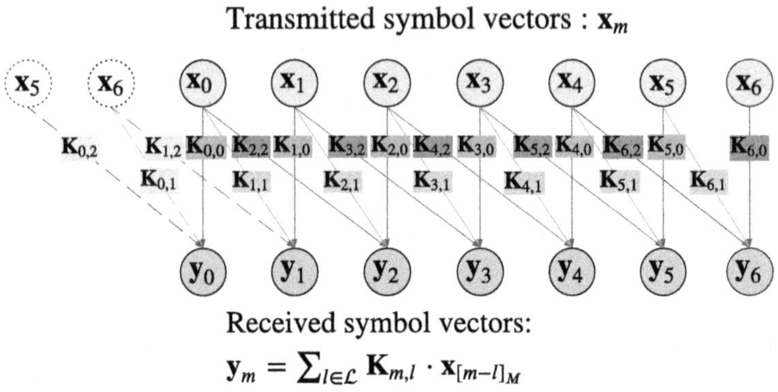

FIGURE 4.20 Factor graph representation of the delay-Doppler domain I/O relation.

$\mathcal{L} = \{0, 1, 2\}$. The figure shows that symbol vectors $\mathbf{x}_{[m-l]_M}$, interfere with each other due to the delay spread, while the components within each symbol vector interfere due to the Doppler shift effect, modeled by $\mathbf{K}_{m,l}$ multiplying each symbol vector.

The primary difference between the OTFS variants lies in the structures of the delay-time and delay-Doppler channel matrices $\tilde{\mathbf{H}}$ and $\mathbf{H}$: when $m < l$, for CP/ZP-OTFS, they are lower block triangular matrices, while for RCP/RZP-OTFS, they have nonzero blocks in the top-right corner (see examples in Figs. 4.14 and 4.15). Table 4.2 summarizes the different submatrices $\tilde{\mathbf{K}}_{m,l}$ and $\mathbf{K}_{m,l}$ for $m \geq l$ and $m < l$. For example, in Fig. 4.20, $\mathbf{K}_{0,1}$, $\mathbf{K}_{0,2}$, and $\mathbf{K}_{1,2}$ (gray-shaded) are for $m < l$, while the remaining are for $m \geq l$.

The circulant property of these submatrices $\mathbf{K}_{m,l}$ can be utilized to design efficient detectors for OTFS to be discussed in Chapter 6. Another desirable property of channel matrix $\mathbf{H}$ is its sparsity, which determines the detection complexity. Let S be the number of nonzero elements in each row of the channel matrix $\mathbf{H}$. Since we are operating in the discrete delay-Doppler domain, S corresponds to the number of observed delay-Doppler paths in the delay-Doppler grid and should not be mistaken with the actual number of propagation paths P. The grid samples can denote only integer delay and Doppler taps. However, in practical cases, multipath channels with P paths may result in $S \geq P$ due to leakage caused by fractional delay and Doppler paths as explained in Section 4.5 and illustrated in the case of fractional Doppler in Fig. 4.16. The number of discrete paths S can be brought closer to P for a constant $M\Delta f$ by increasing the duration NT or by changing the pulse shaping waveform $g_{\text{tx}}(t)$.

TABLE 4.2 Submatrices $\tilde{\mathbf{K}}_{m,l}$ and $\mathbf{K}_{m,l}$ for variants of OTFS.

	$m \geq l$	$m < l$
CP	$\tilde{\mathbf{K}}_{m,l} = \text{diag}[\tilde{\boldsymbol{\nu}}_{m,l}]$ $\mathbf{K}_{m,l} = \text{circ}[\boldsymbol{\nu}_{m,l}]$	$\tilde{\mathbf{K}}_{m,l} = \text{diag}[\tilde{\boldsymbol{\nu}}_{m,l}]$ $\mathbf{K}_{m,l} = \text{circ}[\boldsymbol{\nu}_{m,l}]$
ZP	$\tilde{\mathbf{K}}_{m,l} = \text{diag}[\tilde{\boldsymbol{\nu}}_{m,l}]$ $\mathbf{K}_{m,l} = \text{circ}[\boldsymbol{\nu}_{m,l}]$	$\tilde{\mathbf{K}}_{m,l} = \mathbf{0}_{N,N}$ $\mathbf{K}_{m,l} = \mathbf{0}_{N,N}$
RCP	$\tilde{\mathbf{K}}_{m,l} = \text{diag}[\tilde{\boldsymbol{\nu}}_{m,l}]$ $\mathbf{K}_{m,l} = \text{circ}[\boldsymbol{\nu}_{m,l}]$	$\tilde{\mathbf{K}}_{m,l} = \text{diag}[\tilde{\boldsymbol{\nu}}_{m,l}] \cdot \boldsymbol{\Pi}$ $\mathbf{K}_{m,l} = \text{circ}[\boldsymbol{\nu}_{m,l}] \cdot \mathbf{D}$
RZP	$\tilde{\mathbf{K}}_{m,l} = \text{diag}[\tilde{\boldsymbol{\nu}}_{m,l}]$ $\mathbf{K}_{m,l} = \text{circ}[\boldsymbol{\nu}_{m,l}]$	$\tilde{\mathbf{K}}_{m,l} = \text{diag}[0, \tilde{\nu}_{m,l}(1), ...\tilde{\nu}_{m,l}(N-1)] \cdot \boldsymbol{\Pi}$ $\mathbf{K}_{m,l} = (\text{circ}[\boldsymbol{\nu}_{m,l}] - \frac{\tilde{\nu}_{m,l}(0)}{N}\mathbf{I}_{N,N}) \cdot \mathbf{D}$

4.6.2 Delay-Doppler input–output relations for OTFS variants

Consider a wireless channel with P paths, where each has channel gain g_i with the normalized delay shift ℓ_i and Doppler shift κ_i, for $i = 1, \ldots, P$. When the receiver has sufficient delay and Doppler resolutions, fractional delay and Doppler shifts do not cause significant leakage into the neighboring delay and Doppler grid points, respectively. Under such assumption, we can assume that $\ell_i = l_i$ and $\kappa_i = k_i$ are integers. In the case of CP/ZP-OTFS (see Section 4.5) we assume $\kappa\gamma_g = k_i$ are integers. Recall from Section 4.1.3 that the Doppler response $\nu_l(k)$ is

$$\nu_l(k) = \begin{cases} g_i, & \text{if } l = l_i \text{ and } k = k_i, \\ 0, & \text{otherwise.} \end{cases} \tag{4.116}$$

We now make a summary of the symbol-wise delay-Doppler input–output relations for OTFS variants. We start with CP-OTFS, since it has the simplest input–output expression among all.

1. **CP-OTFS**: Based on (4.101) and (4.107), after removing CP, the delay-Doppler input–output relation is

$$\begin{aligned} \mathbf{y}_m[n] &= \sum_{l\in\mathcal{L}} \sum_{k=0}^{N-1} \boldsymbol{\nu}_{m,l}[k]\mathbf{x}_{[m-l]_M}\big[[n-k]_N\big] \\ &= \sum_{l\in\mathcal{L}} \sum_{k=-N/2}^{N/2-1} \nu_l(k) z^{k(m-l)} \mathbf{x}_{[m-l]_M}\big[[n-k]_N\big]. \end{aligned} \tag{4.117}$$

Replacing $\mathbf{y}_m[n]$ with $\mathbf{Y}[m,n]$ and $\mathbf{x}_m[n]$ with $\mathbf{X}[m,n]$ and substituting (4.116) in (4.117), we obtain, for both $m \geq l_i$ and $m < l_i$,

$$\mathbf{Y}[m,n] = \sum_{i=1}^{P} g_i z^{k_i(m-l_i)} \mathbf{X}\big[[m-l_i]_M, [n-k_i]_N\big], \tag{4.118}$$

where $m = 0, \ldots, M-1$ and $n = 0, \ldots, N-1$. It can be noted that the received information symbol $\mathbf{Y}[m,n]$ is the output of 2D convolution of the transmitted symbol $\mathbf{X}[m,n]$ with a filter with time-varying coefficients due to phase rotations $z^{k_i(m-l_i)}$. This is different from the ideal pulse shaping case in (4.15), where the delay-Doppler domain received symbol is the output of a time-invariant 2D circular convolution.

2. **ZP-OTFS**: The only difference from the previous case is that there is no circular interference along the delay domain. After discarding ZP, the input–output relation in (4.118) can be simply modified by removing the modulo-M operation from the first index of $\mathbf{X}$ as

$$\mathbf{Y}[m,n] = \sum_{i=1}^{P} g_i z^{k_i(m-l_i)} \mathbf{X}\big[m - l_i, [n-k_i]_N\big], \quad m \geq l_i, \tag{4.119}$$

where $\mathbf{X}[m-l_i, n] = 0$ for $m < l_i$.

3. **RCP-OTFS**: For $m \geq l_i$, the input–output relation can be written as

$$\mathbf{Y}[m,n] = \sum_{i=1}^{P} g_i z^{k_i(m-l_i)} \mathbf{X}\big[m - l_i, [n-k_i]_N\big], \quad m \geq l_i. \tag{4.120}$$

For $m < l_i$, we have

$$\mathbf{Y}[m,n] = \sum_{i=1}^{P} g_i z^{k_i([m-l_i]_M)} \mathrm{e}^{-j2\pi n/N} \mathbf{X}\big[[m - l_i]_M, [n-k_i]_N\big], \quad m < l_i. \tag{4.121}$$

4. **RZP-OTFS**: For $m \geq l_i$, the input–output relation in (4.120) is also valid to RZP-OTFS. When $m < l_i$, from Section 4.5.1, the RZP-OTFS input–output relation does not have a compact form as the other cases, but can be approximated for large N as

$$\mathbf{Y}[m,n] \approx \sum_{i=1}^{P} g_i \left(\frac{N-1}{N}\right) z^{k_i(m-l_i)} \mathrm{e}^{\frac{-j2\pi}{N}[n-k_i]_N} \mathbf{X}\big[[m - l_i]_M, [n-k_i]_N\big]. \tag{4.122}$$

The detailed steps of the above approximation can be found in Chapter 5 deriving the RZP-OTFS input–output relation using the Zak transform.

Assuming CP/ZP removal in all OTFS variants, we can summarize a general form covering all the above cases as

$$\boxed{\mathbf{Y}[m,n] = \sum_{i=1}^{P} g_i \alpha_{l_i,k_i}[m,n] \mathbf{X}\big[[m - l_i]_M, [n-k_i]_N\big],} \tag{4.123}$$

where the phase rotations $\alpha_{l_i,k_i}[m,n]$ can be summarized as shown in Table 4.3.

TABLE 4.3 Phase rotations $\alpha_{l_i,k_i}[m,n]$ for variants of OTFS.

	$m \geq l_i$	$m < l_i$
CP	$z^{k_i(m-l_i)}$	$z^{k_i(m-l_i)}$
ZP	$z^{k_i(m-l_i)}$	0
RCP	$z^{k_i(m-l_i)}$	$z^{k_i([m-l_i]_M)}e^{-j2\pi n/N}$
RZP	$z^{k_i(m-l_i)}$	$(\frac{N-1}{N})z^{k_i(m-l_i)}e^{\frac{-j2\pi}{N}[n-k_i]_N}$

4.6.3 Comparison of OTFS variants

For the OTFS variants discussed in the chapter, Table 4.4 compares them based on the normalized spectral efficiency (NSE),[3] the total transmit power (P_T), and the sparsity of the channel matrix **H**. The sparsity of the channel matrix **H** provides a coarse indication of the complexity of channel estimation and the MP detection.

Consider Q-QAM modulation alphabets with $\log_2 Q$ bits per symbol and an average symbol energy E_s. Table 4.4 shows that RZP/RCP-OTFS offer higher NSE thanks to a CP/ZP per frame, while ZP/CP-OTFS have lower NSE due to a CP/ZP per block. In all cases, a large value of M much greater than L_{CP} or L_{ZP} guarantees a high NSE.

In terms of the transmitted power (P_T), ZP/RZP-OTFS requires the least power, while CP-OTFS requires more power, since L_{CP} additional guard samples are transmitted per block.

We study the sparsity of matrix **H** based on channel submatrices $\mathbf{K}_{m,l}$ in Table 4.2. It can be observed that ZP-OTFS has the sparsest channel matrix, since the top-right corner submatrices ($\mathbf{K}_{m,l}$ for $m < l$) are zero matrices. Both CP/RCP-OTFS have the same channel sparsity, higher than ZP-OTFS. For RZP-OTFS, the top-right corner submatrices ($\mathbf{K}_{m,l}$, for $m < l$) are full matrices (4.76), thereby making **H** the least sparse. However, for large N, the RZP-OTFS input–output relation can be approximated as in (4.122) and as a result, its sparsity approaches that of CP/RCP-OTFS.

The choice among OTFS variants depends on not only the design constraints discussed here, but also the complexity of detection and channel estimation, to be discussed in Chapters 6 and 7.

[3]Here, normalized spectral efficiency is the ratio between the original spectral efficiency and $\log_2 Q$ for Q-QAM signalling.

TABLE 4.4 Comparison of OTFS variants.

	Normalized spectral eff.	Transmit power P_T	Sparsity of H
CP	$\frac{M}{M+L_{CP}}$	$N(M+L_{CP})E_s$	low
ZP	$\frac{M}{M+L_{ZP}}$	NME_s	lowest
RCP	$\frac{NM}{NM+L_{CP}}$	$(NM+L_{CP})E_s$	low
RZP	$\frac{NM}{NM+L_{ZP}}$	NME_s	moderate

4.7 Bibliographical notes

OTFS modulation was first presented by Hadani et al. at the 2017 IEEE Wireless Communications and Networking Conference [1]. The OTFS input–output relations were explicitly derived for RZP/RCP-OTFS in [2–4], where the approximation in (4.122) has been detailed in [2]. Then the OTFS input–output relations in matrix form for RZP/RCP-OTFS were presented in [5–7], and the relation for ZP/CP-OTFS was presented in [8–16]. OTFS modulation was derived using Zak transform principles in [17,18]. OTFS was generalized in the form of 2D orthogonal precoding in the time-frequency domain in [19]. An error performance analysis was conducted for OTFS in delay-Doppler domain in [20,21]. It can be also noted that the OTFS modulation is similar to asymmetric OFDM proposed for static wireless channels [22]. The connections of OTFS modulation to other modulation techniques were explored in [14,23,24].

References

[1] R. Hadani, S. Rakib, M. Tsatsanis, A. Monk, A.J. Goldsmith, A.F. Molisch, R. Calderbank, Orthogonal time frequency space modulation, in: 2017 IEEE Wireless Communications and Networking Conference (WCNC), 2017, pp. 1–6.

[2] P. Raviteja, K.T. Phan, Y. Hong, E. Viterbo, Interference cancellation and iterative detection for orthogonal time frequency space modulation, IEEE Transactions on Wireless Communications 17 (10) (2018) 6501–6515, https://doi.org/10.1109/TWC.2018.2860011.

[3] K.R. Murali, A. Chockalingam, On OTFS modulation for high-Doppler fading channels, in: 2018 Information Theory and Applications Workshop (ITA), 2018, pp. 1–6.

[4] L. Gaudio, M. Kobayashi, G. Caire, G. Colavolpe, On the effectiveness of OTFS for joint radar parameter estimation and communication, IEEE Transactions on Wireless Communications 19 (9) (2020) 5951–5965, https://doi.org/10.1109/TWC.2020.2998583.

[5] P. Raviteja, Y. Hong, E. Viterbo, E. Biglieri, Practical pulse-shaping waveforms for reduced-cyclic-prefix OTFS, IEEE Transactions on Vehicular Technology 68 (1) (2019) 957–961, https://doi.org/10.1109/TVT.2018.2878891.

[6] P. Raviteja, Y. Hong, E. Viterbo, E. Biglieri, Effective diversity of OTFS modulation, IEEE Communications Letters 9 (2) (2020) 249–253, https://doi.org/10.1109/LWC.2019.2951758.

[7] S. Tiwari, S.S. Das, Circularly pulse-shaped orthogonal time frequency space modulation, Electronics Letters 56 (3) (2020) 157–160, https://doi.org/10.1049/el.2019.2503.

[8] A. Farhang, A. RezazadehReyhani, L.E. Doyle, B. Farhang-Boroujeny, Low complexity modem structure for OFDM-based orthogonal time frequency space modulation, IEEE

Wireless Communications Letters 7 (3) (2018) 344–347, https://doi.org/10.1109/LWC.2017.2776942.

[9] A. Rezazadehreyhani, A. Farhang, A. Ji, R.R. Chen, B. Farhang-Boroujeny, Analysis of discrete-time MIMO OFDM-based orthogonal time frequency space modulation, in: 2018 IEEE International Conference on Communications, 2018, pp. 1–6.

[10] W. Shen, L. Dai, J. An, P.Z. Fan, R.W. Heath, Channel estimation for orthogonal time frequency space (OTFS) massive MIMO, IEEE Transactions on Signal Processing 67 (16) (2019) 4204–4217, https://doi.org/10.1109/TSP.2019.2919411.

[11] T. Thaj, E. Viterbo, Low complexity iterative rake decision feedback equalizer for zero-padded OTFS systems, IEEE Transactions on Vehicular Technology 69 (12) (2020) 15606–15622, https://doi.org/10.1109/TVT.2020.3044276.

[12] T. Thaj, E. Viterbo, Y. Hong, Orthogonal time sequency multiplexing modulation: analysis and low-complexity receiver design, IEEE Transactions on Wireless Communications 20 (12) (2021) 7842–7855, https://doi.org/10.1109/TWC.2021.3088479.

[13] M.K. Ramachandran, G.D. Surabhi, A. Chockalingam, OTFS: a new modulation scheme for high-mobility use cases, Journal of the Indian Institute of Science (2020) 315–336, https://doi.org/10.1007/s41745-020-00167-4.

[14] V. Rangamgari, S. Tiwari, S.S. Das, S.C. Mondal, OTFS: interleaved OFDM with block CP, in: 2020 IEEE National Conference on Communications (NCC), 2020, pp. 1–6.

[15] S.S. Das, V. Rangamgari, S. Tiwari, S.C. Mondal, Time domain channel estimation and equalization of CP-OTFS under multiple fractional Dopplers and residual synchronization errors, IEEE Access 9 (2020) 10561–10576, https://doi.org/10.1109/ACCESS.2020.3046487.

[16] D. Shi, W. Wang, L. You, X. Song, Y. Hong, X. Gao, G. Fettweis, Deterministic pilot design and channel estimation for downlink massive MIMO-OTFS systems in presence of the fractional Doppler, IEEE Transactions on Wireless Communications 20 (11) (2021) 7151–7165, https://doi.org/10.1109/TWC.2021.3081164.

[17] S.K. Mohammed, Derivation of OTFS modulation from first principles, IEEE Transactions on Vehicular Technology 70 (8) (2021) 7619–7636, https://doi.org/10.1109/TVT.2021.3069913.

[18] S.K. Mohammed, Time-domain to delay-Doppler domain conversion of OTFS signals in very high mobility scenarios, IEEE Transactions on Vehicular Technology 70 (6) (2021) 6178–6183, https://doi.org/10.1109/TVT.2021.3071942.

[19] T. Zemen, M. Hofer, D. Löschenbrand, C. Pacher, Iterative detection for orthogonal precoding in doubly selective channels, in: 2018 IEEE 29th Annual International Symposium on Personal, Indoor and Mobile Radio Communications (PIMRC), 2018, pp. 1–7.

[20] E. Biglieri, P. Raviteja, Y. Hong, Error performance of orthogonal time frequency space (OTFS) modulation, in: 2019 IEEE International Conference on Communications Workshops (ICC Workshops), 2019, pp. 1–6.

[21] Z. Wei, W. Yuan, S. Li, J. Yuan, D.W.K. Ng, Transmitter and receiver window designs for orthogonal time-frequency space modulation, IEEE Transactions on Communications 69 (4) (2021) 2207–2223, https://doi.org/10.1109/TCOMM.2021.3051386.

[22] J. Zhang, A.D.S. Jayalath, Y. Chen, Asymmetric OFDM systems based on layered FFT structure, IEEE Signal Processing Letters 14 (11) (2007) 812–815, https://doi.org/10.1109/LSP.2007.903230.

[23] A. Nimr, M. Chafii, M. Matthe, G. Fettweis, Extended GFDM framework: OTFS and GFDM comparison, in: 2018 IEEE Global Communications Conference (GLOBECOM), 2018, pp. 1–6.

[24] G.D. Surabhi, R.M. Augustine, A. Chockalingam, Peak-to-average power ratio of OTFS modulation, IEEE Commununications Letters 23 (6) (2019) 999–1002, https://doi.org/10.1109/LCOMM.2019.2914042.

CHAPTER

5

Zak transform analysis for delay-Doppler communications

OUTLINE

Delay-Doppler Communications
https://doi.org/10.1016/B978-0-32-385028-5.00013-X

Chapter points

- Basic properties of the Zak transform.
- Continuous Zak transform in delay-Doppler communications.
- Discrete Zak transform in delay-Doppler communications.

The butterfly counts not months but moments, and has time enough.
Rabindranath Tagore

This chapter reviews the fundamental concepts of delay-Doppler analysis and processing of signals through the Zak transform (ZT). Some basic facts about the different types of Fourier transform (for continuous- and discrete-time signals) are first reviewed, and the interpretation in terms of harmonic analysis is explicitly presented by listing the different types of orthogonal Fourier basis functions. Then the Zak transform is presented with all its properties and corresponding basis functions. Finally, the delay-Doppler input–output relation for transmission of a signal through a high mobility channel is made explicit in terms of Zak transforms.

5.1 A brief review of the different Fourier transforms

The frequency spectrum of time domain signals is provided by the four different forms of the Fourier transform, depending on the types of time domain signals (discrete/continuous, periodic/nonperiodic). Table 5.1 illustrates how the Fourier transform is a function of a frequency variable, which can be a discrete index k or a continuous variable f. For discrete-time signals, the spectrum is periodic, and it is sufficient to consider a limited range of frequencies covering one period. For continuous-time signals, the frequency axis is either discrete (in $\mathbb{Z}$) or continuous (in $\mathbb{R}$). We refer to the domain of the Fourier transforms as the *frequency domain*, and we remark that it is represented on a one-dimensional axis.

In harmonic analysis, a time domain signal is projected (by taking a suitably defined scalar product) on orthonormal Fourier basis functions, which are periodic time signals. The inverse of the period of such Fourier basis functions defines the point on the frequency axis for which the spectrum of a time domain signal is evaluated. A time domain signal can be synthesized as a combination of Fourier basis signals as summarized in Table 5.2.

TABLE 5.1 The four types of Fourier transforms with different frequency domains.

	Discrete-time (n)	**Continuous-time (t)**
Time periodic by N or T	DFT (k) $0 \le k < N$	Fourier series (k) $-\infty < k < +\infty$
Time nonperiodic	DTFT (f) $0 \le f < 1$ Periodic in frequency	CTFT (f) $-\infty < f < +\infty$ Nonperiodic in frequency

TABLE 5.2 The periodic Fourier basis functions of the different Fourier transforms.

DFT $p_k[n] = \exp(j\frac{2\pi k}{N}n)$	$x[n] = \frac{1}{\sqrt{N}}\sum_{k=0}^{N-1} X[k]p_k[n]$ $X[k] = \langle x[k], p_k[n]\rangle \triangleq \frac{1}{\sqrt{N}}\sum_{n=0}^{N-1} x[n]p_k^*[n]$
Fourier series (period T) $\phi_k(t) = \exp(j\frac{2\pi k}{T}t)$	$x(t) = \sum_{k=-\infty}^{+\infty} a_k\phi_k(t)$ $a_k = \langle x(t), \phi_k(t)\rangle \triangleq \frac{1}{T}\int_0^T x(t)\phi_k^*(t)dt$
DTFT (sampling period T_s) $\psi_n(f) = \exp(j2\pi nT_s f)$	$x[n] = x(nT_s) = \sqrt{T_s}\int_0^{1/T_s} X(f)\psi_n(f)df$ $X(f) = \langle x[n], \psi_n(f)\rangle \triangleq \sqrt{T_s}\sum_{n=-\infty}^{+\infty} x[n]\psi_n^*(f)$
CTFT $\psi(f,t) = \exp(j2\pi ft)$	$x(t) = \int_{-\infty}^{+\infty} X(f)\psi(f,t)df$ $X(f) = \langle x(t), \psi(f,t)\rangle \triangleq \int_{-\infty}^{+\infty} x(t)\psi^*(f,t)dt$

In other words, the Fourier transform provides a representation of time domain signals in a different coordinate system defined by the set of orthonormal Fourier basis functions. Fig. 5.1 illustrates how an abstract signal, represented by the vector ending in the black dot, can be represented in two different coordinate systems: **x** defined by the time basis and **X** defined by the Fourier basis.

The coordinate axis transformation is unitary (or orthogonal), i.e., it preserves distances between any pair of points. This property underpins the well-known Parseval identity:

$$\underbrace{\langle x, y\rangle}_{\text{time}} = \underbrace{\langle \mathcal{F}(x), \mathcal{F}(y)\rangle}_{\text{frequency}}, \tag{5.1}$$

which means that the angle between any pair of vectors is independent of the coordinate system, be it the time basis or the Fourier basis. Considering the physical interpretation, the scalar product defines the correlation between two signals in time domain and the scalar product of a signal with itself defines its energy (or power for periodic signals). As a consequence of (5.1), the energy or power of a time domain signal can be equivalently obtained from its representation in the frequency domain.

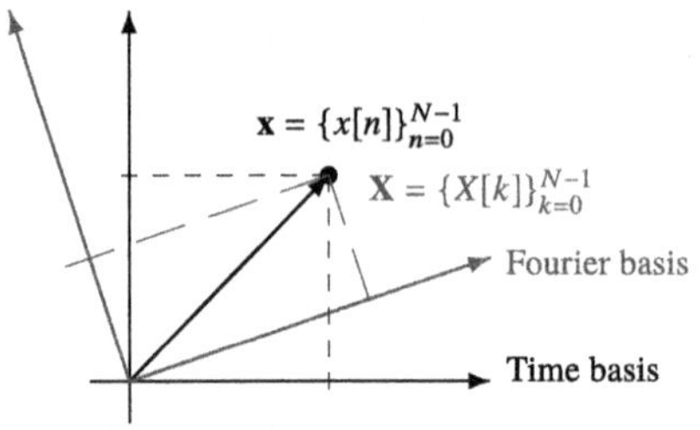

FIGURE 5.1 The Fourier transform viewed as a change of coordinate system.

In conclusion, given the periodic behavior the Fourier basis functions, the Fourier transform highlights the spectral content of a given time domain signal at each frequency. As commonly taught in basic signal processing, the frequency variable can be discrete or continuous and the frequency spectrum is a function over the one-dimensional frequency domain. We will see in the following how expanding the transform domain to a two-dimensional plane enables to fundamentally analyze different characteristics of signals beyond their frequency content.

5.2 The Zak transform

The Zak transform provides a powerful tool for delay-Doppler domain signal analysis and processing. Given a continuous-time signal $x(t)$, we define the *Zak transform* with step T as the bivariate function over the delay-Doppler (τ, ν) plane

$$Z_T\big[x(t)\big](\tau,\nu) = \sqrt{T}\sum_{n=-\infty}^{+\infty} x(\tau+nT)e^{-j2\pi nT\nu}, \quad -\infty < \tau, \nu < \infty. \tag{5.2}$$

We note that for any given time shift (or delay) $-\infty < \tau < \infty$, the sum in (5.2) is the discrete-time Fourier transform (DTFT) of the discrete-time signal $x_\tau[n] = x(\tau + nT)$ obtained by sampling $x(t)$ every T seconds starting at τ. This implies that the Zak transform is periodic in the variable ν, with a period equal to $\Delta f = 1/T$, given the sampling period T_s is equal to T.

In the special case where $\nu = 0$ (no Doppler),

$$Z_T\big[x(t)\big](\tau,0) = \sqrt{T}\sum_{n=-\infty}^{+\infty} x(\tau+nT), \quad -\infty < \tau < \infty, \tag{5.3}$$

the Zak transform is also periodic in τ with period T and coincides with the periodic replica of $x(t)$.

Similarly to the Fourier transform, the Zak transform is the result of a projection of the time domain signal $x(t)$ onto the time domain Zak basis functions parametrized by τ and ν

$$Z_T\big[x(t)\big](\tau,\nu) = \langle x(t), \Phi_{\tau,\nu}(t)\rangle = \int_{-\infty}^{+\infty} x(t)\Phi_{\tau,\nu}(t)^* dt, \tag{5.4}$$

where $(\cdot)^*$ denotes the complex conjugation and

$$\Phi_{\tau,\nu}(t) = \sqrt{T}\sum_{n=-\infty}^{+\infty}\delta(t-\tau-nT)e^{j2\pi(t-\tau)\nu} = \sqrt{T}\sum_{n=-\infty}^{+\infty}\delta(t-\tau-nT)e^{j2\pi nT\nu}. \tag{5.5}$$

These Zak basis functions are pulse trains shifted by τ and modulated by a complex sine wave at frequency ν. Hadani, the inventor of OTFS, has referred to these functions as *pulsones* since they can be visualized as a *pulse* train modulated by a complex *tone*.

5.2.1 Properties of the Zak transform

The following properties of the Zak transform can be easily proved.

1. Linearity:

$$Z_T\big[ax(t)+by(t)\big](\tau,\nu) = aZ_T\big[x(t)\big](\tau,\nu) + bZ_T\big[y(t)\big](\tau,\nu). \tag{5.6}$$

2. The Zak transform is periodic in ν with period $\Delta f = 1/T$:

$$Z_T\big[x(t)\big](\tau,\nu+\Delta f) = Z_T\big[x(t)\big](\tau,\nu). \tag{5.7}$$

3. Translation in time by $0 \le \tau_0 < T$:

$$Z_T\big[x(t+\tau_0)\big](\tau,\nu) = Z_T\big[x(t)\big](\tau+\tau_0,\nu). \tag{5.8}$$

If $\tau_0 = mT$, an integer multiple of T, then

$$Z_T\big[x(t+mT)\big](\tau,\nu) = e^{j2\pi mT\nu} Z_T\big[x(t)\big](\tau,\nu). \tag{5.9}$$

We say the Zak transform is *quasiperiodic* in τ with period T, i.e., it is periodic up to a phase shift $e^{j2\pi mT\nu}$ for each period interval $[mT,(m+1)T)$.

4. Modulation by a complex sine wave with frequency ν_0:

$$Z_T\big[e^{j2\pi\nu_0 t}x(t)\big](\tau,\nu) = e^{j2\pi\nu_0\tau} Z_T\big[x(t)\big](\tau,\nu-\nu_0). \tag{5.10}$$

If $\nu_0 = m\Delta f = m/T$, an integer multiple of Δf, then

$$Z_T\big[e^{j2\pi m\Delta f t}x(t)\big](\tau,\nu) = Z_T\big[x(t)\big](\tau,\nu). \tag{5.11}$$

5. Joint translation by τ_0 and modulation by a complex sine wave with frequency ν_0:

$$Z_T\left[e^{j2\pi\nu_0 t}x(t-\tau_0)\right](\tau,\nu) = e^{j2\pi\nu_0(\tau-\tau_0)}Z_T\left[x(t)\right](\tau-\tau_0,\nu-\nu_0). \quad (5.12)$$

6. Conjugation:

$$Z_T\left[x^*(t)\right](\tau,\nu) = Z_T^*\left[x(t)\right](\tau,-\nu). \quad (5.13)$$

7. Symmetry:

$$Z_T\left[x(t)\right](\tau,\nu) = Z_T^*\left[x(t)\right](-\tau,-\nu) \qquad \text{if } x(t) \text{ even}, \quad (5.14)$$

$$Z_T\left[x(t)\right](\tau,\nu) = -Z_T^*\left[x(t)\right](-\tau,-\nu) \qquad \text{if } x(t) \text{ odd}. \quad (5.15)$$

8. Product in time:

$$y(t) = h(t)x(t). \quad (5.16)$$

Then

$$Z_T\left[y(t)\right](\tau,\nu) = \sqrt{T}\int_0^{\Delta f} Z_T\left[h(t)\right](\tau,u)Z_T\left[x(t)\right](\tau,u-\nu)\,du. \quad (5.17)$$

This can be interpreted as a convolution in the Doppler domain for any given delay τ.

9. Convolution in time:

$$y(t) = h(t) * x(t) = \int_{-\infty}^{+\infty} h(\theta)x(t-\theta)\,d\theta. \quad (5.18)$$

Then

$$Z_T\left[y(t)\right](\tau,\nu) = \frac{1}{\sqrt{T}}\int_0^{T} Z_T\left[h(t)\right](\theta,\nu)Z_T\left[x(t)\right](\tau-\theta,\nu)\,d\theta. \quad (5.19)$$

This can be interpreted as a convolution in the delay domain for any given Doppler shift ν.

A sample proof. As an example to familiarize the reader with this tool, we show the step-by-step proof of (5.19):

$$\begin{aligned} Z_T\left[y(t)\right](\tau,\nu) &= \sqrt{T}\sum_n y(\tau+nT)e^{-j2\pi nT\nu} \\ &= \sqrt{T}\int_{-\infty}^{+\infty} h(\theta)\sum_n x(\tau+nT-\theta)e^{-j2\pi nT\nu}\,d\theta \\ &= \int_{-\infty}^{+\infty} h(\theta)\,Z_T\left[x(t)\right](\tau-\theta,\nu)\,d\theta \end{aligned}$$

$$
\begin{aligned}
&= \sum_n \int_{nT}^{(n+1)T} h(\theta) Z_T\big[x(t)\big](\tau - \theta, \nu)\, d\theta \\
&= \sum_n \int_0^T h(\theta + nT) Z_T\big[x(t)\big](\tau - \theta - nT, \nu)\, d\theta \\
&\overset{(a)}{=} \int_0^T \sum_n h(\theta + nT) e^{-j2\pi nT\nu} Z_T\big[x(t)\big](\tau - \theta, \nu)\, d\theta \\
&= \frac{1}{\sqrt{T}} \int_0^T Z_T\big[h(t)\big](\theta, \nu) Z_T\big[x(t)\big](\tau - \theta, \nu)\, d\theta. \qquad (5.20)
\end{aligned}
$$

In step (a) the translation by an integer multiple of T property (5.9) is used. □

Given the periodicity in ν in (5.7) and the quasiperiodicity in τ up to the phase term $e^{j2\pi m\nu}$ in (5.9), the Zak transform is completely determined by its values on the rectangular region

$$\mathcal{R} = \big\{\tau \in [0, T), \nu \in [0, 1/T)\big\}. \qquad (5.21)$$

We refer to $\mathcal{R}$ as the *fundamental region* and denote the restriction of the Zak transform to $\mathcal{R}$ by $\bar{Z}_T[x(t)](\tau, \nu)$, with $\bar{Z}_T[x(t)](\tau, \nu) = 0$ for $(\tau, \nu) \notin \mathcal{R}$. Then we can write

$$
\begin{aligned}
Z_T\big[x(t)\big](\tau, \nu) &= \sum_n \sum_m \bar{Z}_T\big[x(t)\big](\tau - nT, \nu - m\Delta f) e^{j2\pi nT(\nu - m\Delta f)} \\
&= \sum_n \sum_m \bar{Z}_T\big[x(t)\big](\tau - nT, \nu - m\Delta f) e^{j2\pi nT\nu}. \qquad (5.22)
\end{aligned}
$$

Fig. 5.2 provides a graphical description of (5.22), showing the periodicity along Doppler and the quasiperiodicity along delay.

5.2.2 The inverse Zak transform

The inverse Zak transform returns the time domain signal $x(t)$ as

$$x(t) = \sqrt{T} \int_0^{\Delta f} Z_T\big[x(t)\big](\tau, \nu) d\nu, \quad -\infty < \tau < \infty. \qquad (5.23)$$

In order to clarify how (5.23) operates, we substitute (5.22) and use the properties (5.9) and (5.11) to obtain

$$x(t) = \sqrt{T} \sum_{n=-\infty}^{\infty} \int_0^{\Delta f} \bar{Z}_T\big[x(t)\big](\tau + nT, \nu) e^{j2\pi nT\nu} d\nu = \sum_{n=-\infty}^{\infty} x(\tau + nT), \qquad (5.24)$$

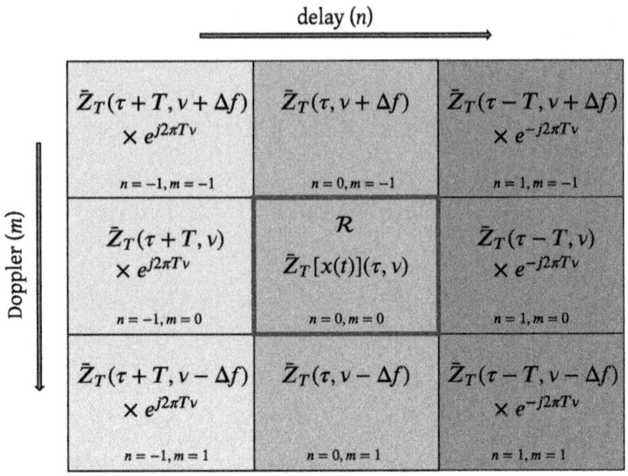

FIGURE 5.2 Periodicity and quasiperiodicity of the Zak transform.

where $0 \le \tau < T$ and $-\infty < t < \infty$ is the time variable obtained by concatenating all the intervals $[nT, (n+1)T)$. For any given time shift (delay) $0 \le \tau < T$, the integral in (5.24) is an inverse discrete-time Fourier transform (IDTFT) that yields the discrete-time signal samples $x_\tau[n] = x(\tau + nT)$ obtained by sampling $x(t)$ every T seconds starting at τ. When all the interleaved samples are concatenated, $x(t)$ is formed. This reconstruction operation will become more natural with the discrete Zak transform (DZT) discussed in Section 5.5.

5.3 The delay-Doppler basis functions

As discussed before, the delay-Doppler basis functions $\Phi_{\tau_0,\nu_0}(t)$ are time domain signals parametrized by the delay and Doppler variables τ_0 and ν_0. They can also be found as inverse Zak transforms of a Dirac pulse at $(\tau_0, \nu_0) \in \mathcal{R}$, i.e.,

$$\bar{Z}_T\left[\Phi_{\tau_0,\nu_0}(t)\right](\tau,\nu) = \delta(\tau-\tau_0)\delta(\nu-\nu_0), \tag{5.25}$$

resulting in

$$\Phi_{\tau_0,\nu_0}(t) = \sqrt{T}\sum_{n=-\infty}^{\infty} e^{j2\pi nT\nu_0}\delta(t-\tau_0-nT). \tag{5.26}$$

The *"DC" basis function* for $\tau_0 = 0$ and $\nu_0 = 0$ is the pulse train

$$\Phi_{0,0}(t) = \sqrt{T} \sum_{n=-\infty}^{\infty} \delta(t - nT). \tag{5.27}$$

The basis function with only delay τ_0 and $\nu_0 = 0$ is the pulse train shifted by τ_0:

$$\Phi_{\tau_0,0}(t) = \sqrt{T} \sum_{n=-\infty}^{\infty} \delta(t - \tau_0 - nT). \tag{5.28}$$

For the general case in (5.26), when $\nu_0 \neq 0$ the pulse train is modulated by a complex sinusoid $e^{j2\pi t \nu_0}$ with a frequency ν_0. Fig. 5.3 shows the magnitude

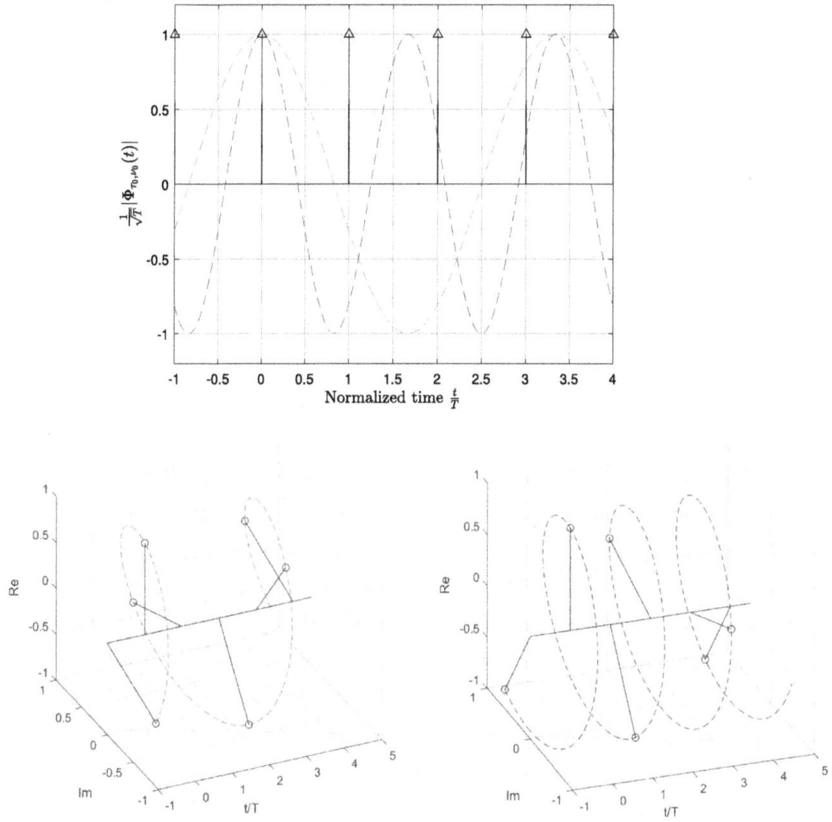

FIGURE 5.3 $\frac{1}{\sqrt{T}}|\Phi_{\tau_0,\nu_0}(t)|$ for normalized delay ($\frac{\tau_0}{T} = 0$) and normalized Doppler shift ($\frac{\nu_0}{\Delta f} = 0.3$) (modulated by the red (light gray in print version) dashed line) and ($\frac{\nu_0}{\Delta f} = 0.6$) (modulated by the blue (dark gray in print version) dashed line).

of the basis functions $\frac{1}{\sqrt{T}}|\Phi_{\tau_0,\nu_0}(t)|$ for normalized delay ($\frac{\tau_0}{T}=0$). The red (light gray in print version) and blue (dark gray in print version) dashed line denotes the real part of the modulating exponential $\mathrm{Re}\{e^{j2\pi\nu_0(t-\tau_0)}\}$ for $\frac{\nu_0}{\Delta f}=0.3$ and $\frac{\nu_0}{\Delta f}=0.6$, respectively. The three-dimensional visualization is also given in the figure for the two Doppler shifts.

The Zak basis signals $\Phi_{\tau_0,\nu_0}(t)$ in the Zak domain correspond to a single delta pulse at (τ_0,ν_0) in the fundamental rectangle $[0,T]\times[0,\Delta f]$ of the Zak domain as shown in (5.25). In this domain it becomes apparent that any pair of delta pulses in distinct locations of the delay-Doppler plane are orthogonal and the set of all these delta pulses for all (τ_0,ν_0) form an orthonormal basis in the delay-Doppler domain. As a consequence, the time domain Zak basis signals $\Phi_{\tau_0,\nu_0}(t)$ also form an orthonormal basis.

5.4 Zak transform in delay-Doppler communications

5.4.1 Single path delay-Doppler channel

If $s(t)$ is transmitted across a single path channel with delay τ_p, Doppler shift ν_p, and gain g_p, then the received signal is

$$\boxed{r(t)=g_p e^{j2\pi\nu_p(t-\tau_p)}s(t-\tau_p)} \tag{5.29}$$

and its corresponding Zak transform is given by

$$\begin{aligned}
Z_T\big[r(t)\big](\tau,\nu) &= \sqrt{T}\sum_n g_p e^{j2\pi\nu_p(\tau+nT-\tau_p)}s(\tau+nT-\tau_p)e^{-j2\pi nT\nu}\\
&= \sqrt{T}\sum_n g_p e^{j2\pi[\nu_p(\tau-\tau_p)-nT(\nu-\nu_p)]}s(\tau+nT-\tau_p)\\
&= g_p e^{j2\pi\nu_p(\tau-\tau_p)}\sqrt{T}\sum_n s(\tau-\tau_p+nT)e^{-j2\pi nT(\nu-\nu_p)}\\
&= g_p e^{j2\pi\nu_p(\tau-\tau_p)}Z_T\big[s(t)\big](\tau-\tau_p,\nu-\nu_p).
\end{aligned} \tag{5.30}$$

The Zak transform of the received signal $r(t)$ is equal to the Zak transform of the transmitted signal $s(t)$, shifted on the two axes by the delay and the Doppler shift of the path and modulated by $e^{j2\pi\nu_p(\tau-\tau_p)}$ along the delay axis.

5.4.2 Multipath and general delay-Doppler channel

Using the linearity property of the Zak transform the result in (5.30) extends trivially to a multipath channel with a sum of P delay-Doppler paths as

$$Z_T\big[r(t)\big](\tau,\nu)=\sum_{p=1}^{P} g_p e^{j2\pi\nu_p(\tau-\tau_p)} Z_T\big[s(t)\big](\tau-\tau_p,\nu-\nu_p). \tag{5.31}$$

In the case where the channel is modeled with the delay-Doppler response $h(\theta,u)$, we replace τ_p and ν_p in (5.31) with the continuous variables θ and u and we replace g_p with the delay-Doppler response $h(\theta,u)=Z_T[h(t)](\theta,u)$. We obtain

$$\begin{aligned} Z_T\big[r(t)\big](\tau,\nu) &= \int_0^T\int_0^{\Delta f} e^{j2\pi u(\tau-\theta)}\, h(\theta,u)\, Z_T\big[s(t)\big](\tau-\theta,\nu-u)\,d\theta\,du \\ &= \int_0^T\int_0^{\Delta f} e^{j2\pi u(\tau-\theta)}\, Z_T\big[h(t)\big](\theta,u)\, Z_T\big[s(t)\big](\tau-\theta,\nu-u)\,d\theta\,du, \end{aligned} \tag{5.32}$$

which corresponds to the two-dimensional twisted convolution of the Zak transforms in the delay-Doppler domain.

Remark. It is important to notice that the delay-Doppler channel applies time shifts and modulations to the input signal and cannot be represented as a linear time-invariant channel, where the output is obtained as the linear convolution of the input with an impulse response. Nevertheless, the input–output relation in the delay-Doppler domain reduces to the twisted two-dimensional convolution (5.31). Such convolution in the delay-Doppler domain is relatively simple given the very sparse representation of the multipath channel in this domain.

If we transmit an information symbol a_0 over a multipath channel using the delay-Doppler domain basis function in (5.26), i.e., $s(t)=a_0\Phi_{\tau_0,\nu_0}(t)$, then we receive

$$r(t)=\sum_{p=1}^{P} a_0 g_p e^{j2\pi\nu_p(t-\tau_p)}\Phi_{\tau_0,\nu_0}(t-\tau_p).$$

Moving to the delay-Doppler domain by taking the Zak transforms we have

$$Z_T\big[s(t)\big](\tau,\nu)=a_0\delta(\tau-\tau_0)\delta(\nu-\nu_0)$$

and

$$Z_T\big[r(t)\big](\tau,\nu)=\sum_{p=1}^{P} a_0 g_p e^{j2\pi\nu_p(\tau-\tau_p)}\delta(\tau-\tau_0+\tau_p)\delta(\nu-\nu_0+\nu_p).$$

The multipath channel spreads any information symbol across P distinct locations inside the relatively small region $[\tau_0, \tau_0 + \tau_{\max}] \times [\nu_0 - \nu_{\max}, \nu_0 + \nu_{\max}]$ of the fundamental rectangle of size $[0, T] \times [0, \Delta f]$ in the delay-Doppler domain. In this case, where a single symbol is transmitted in isolation, a receiver could easily recover all the energy from the P paths. However, when multiplexing multiple information symbols over distinct delay-Doppler domain orthonormal basis signals the scattered symbol locations may overlap and incur intersymbol interference. This indicates that the transmitted orthonormal basis signals marginally lose their orthogonality at the receiver thanks to the localized interference in the delay-Doppler plane caused by the channel.

In the case of CP-OFDM, the information symbols are multiplexed on orthogonal Fourier basis functions, which preserve their orthogonality at the receiver, when transmitted over a static multipath channel with $\tau_{\max}$ shorter than the CP. However, we have seen that delay-Doppler multipath channels cause interference across the entire time-frequency domain. Since the interference is not localized, the loss of orthogonality at the receiver is severe and equalization becomes a more complex task.

5.4.3 Band- and time-limited delay-Doppler basis functions

The Zak basis signals $\Phi_{\tau_0,\nu_0}(t)$ that we defined in previous sections are neither bandlimited nor time limited, since they are essentially pulse trains of infinite duration and infinite bandwidth. In practical communication systems, we deal with approximately band-limited and time-limited signals. It is therefore essential to study how the basis functions look under such limitations.

5.4.3.1 Bandlimited basis functions

Consider an ideal bandlimiting filter with frequency response $H_B(f)$:

$$H_B(f) = \begin{cases} 1 & 0 \le f < M\Delta f, \\ 0 & \text{otherwise.} \end{cases} \tag{5.33}$$

The time domain impulse response of the filter is given by

$$\begin{aligned} h_B(t) &= \int_{-\infty}^{\infty} H_B(f) e^{j2\pi f t} df = \int_0^{M\Delta f} e^{j2\pi f t} df \\ &= e^{j\pi M\Delta f t} M\Delta f \operatorname{sinc}(M\Delta f t). \end{aligned} \tag{5.34}$$

The bandlimited Zak basis functions $\Phi^{B}_{\tau_0,\nu_0}(t)$ are then obtained by convolution with the impulse response of the ideal filter as

$$\boxed{\Phi^{B}_{\tau_0,\nu_0}(t) = \Phi_{\tau_0,\nu_0}(t) * h_B(t).} \tag{5.35}$$

Fig. 5.4 shows the magnitude of basis functions $|\frac{\sqrt{T}}{M}\Phi^{\mathrm{B}}(\tau_0, \nu_0)(t)|$ for normalized delay $\frac{\tau_0}{T} = 0$ and normalized Doppler shift $\frac{\nu_0}{\Delta f} = 0.3$. The dashed line denotes the real part of the phase of the basis functions $\mathrm{Re}\{e^{j2\pi\nu_0(t-\tau_0)}\}$. This can be compared to the magnitude of the corresponding Zak basis function $|\Phi_{\tau_0,\nu_0}(t)|$, without the bandwidth limitation shown in Fig. 5.3.

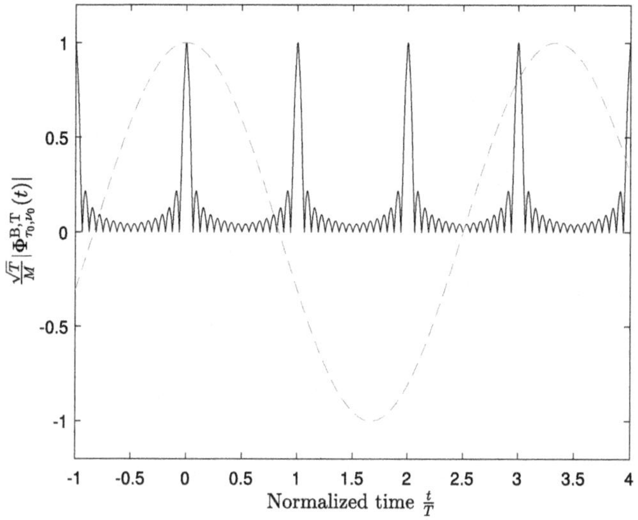

FIGURE 5.4 $|\frac{\sqrt{T}}{M}\Phi^{\mathrm{B,T}}_{\tau_0,\nu_0}(t)|$ for normalized delay ($\frac{\tau_0}{T} = 0$) and normalized Doppler shift ($\frac{\nu_0}{\Delta f} = 0.3$) for $M = 16$.

5.4.3.2 *Time limited basis functions*

The time limited version of the basis functions $\Phi_{\tau_0,\nu_0}(t)$ limited to a duration of NT [s] can be obtained by applying a windowing function h_{T} as

$$\Phi^{\mathrm{T}}_{\tau_0,\nu_0}(t) = \Phi_{\tau_0,\nu_0}(t)h_{\mathrm{T}}(t), \tag{5.36}$$

where

$$h_{\mathrm{T}}(t) = \begin{cases} 1 & 0 \leq t < NT, \\ 0 & \text{otherwise.} \end{cases} \tag{5.37}$$

5.4.3.3 *Band- and time-limited basis functions*

Combining the effects of both band and time limitation, the resultant band- and time-limited (BT) delay-Doppler basis functions $\Phi^{\mathrm{B,T}}_{\tau_0,\nu_0}(t)$ can be written as

$$\boxed{\Phi^{\mathrm{B,T}}_{\tau_0,\nu_0}(t) = \big(\Phi_{\tau_0,\nu_0}(t) * h_{\mathrm{B}}(t)\big)h_{\mathrm{T}}(t).} \tag{5.38}$$

Taking Zak transform on both sides and applying the multiplicative and the convolution property in (5.17) and (5.19), respectively, we obtain

$$\begin{aligned}
Z_T^{\mathrm{B,T}}\left[\Phi_{\tau_0,\nu_0}(t)\right](\tau,\nu) &= Z_T\left[\left(\Phi_{\tau_0,\nu_0}(t) * h_{\mathrm{B}}(t)\right)h_{\mathrm{T}}(t)\right](\tau,\nu)\\
&= \int_{\tau'}\int_{\nu'} \delta\left(\tau'-\tau_0\right)\delta\left(\nu'-\nu_0\right)\cdot\\
&\qquad Z_T\left[h_{\mathrm{B}}(t)\right]\left(\tau-\tau',\nu\right)Z_T\left[h_{\mathrm{T}}(t)\right]\left(\tau',\nu-\nu'\right)d\tau' d\nu'\\
&= Z_T\left[h_{\mathrm{B}}(t)\right](\tau-\tau_0,\nu)\cdot Z_T\left[h_{\mathrm{T}}(t)\right](\tau_0,\nu-\nu_0),
\end{aligned}\tag{5.39}$$

where the Zak transforms of $h_{\mathrm{B}}(t)$ and $h_{\mathrm{T}}(t)$ are given as

$$\boxed{Z_T\left[h_{\mathrm{B}}(t)\right](\tau,\nu) = \frac{1}{\sqrt{T}}e^{j2\pi\nu\tau}e^{-j2\pi\lfloor\frac{\nu}{\Delta f}\rfloor\Delta f\tau}e^{j\pi(M-1)\Delta f\tau}\frac{\sin(\pi M\Delta f\tau)}{\sin(\pi\Delta f\tau)},}\tag{5.40}$$

$$\boxed{Z_T\left[h_{\mathrm{T}}(t)\right](\tau,\nu) = \sqrt{T}e^{j2\pi\nu\lfloor\frac{\tau}{T}\rfloor T}e^{-j\pi\nu(N-1)T}\frac{\sin(\pi\nu NT)}{\sin(\pi\nu T)}.}\tag{5.41}$$

Proof of (5.40). Here we show the derivation of $Z_{\mathrm{T}}[h_{\mathrm{B}}(t)](\tau,\nu)$. From the definition of $h_{\mathrm{B}}(t)$ (in the penultimate step before solving the integral in $[0, M\Delta f]$) in (5.34), its Zak transform can be written as

$$\begin{aligned}
Z_T\left[h_{\mathrm{B}}(t)\right](\tau,\nu) &= \sqrt{T}\sum_n h_{\mathrm{B}}(\tau+nT)e^{-j2\pi nT\nu}\\
&= \sqrt{T}\sum_n\left(\int_0^{M\Delta f} e^{j2\pi f(\tau+nT)}df\right)e^{-j2\pi nT\nu}\\
&= \sqrt{T}\int_0^{M\Delta f} e^{j2\pi f\tau}\left(\sum_n e^{j2\pi(f-\nu)nT}\right)df\\
&= \sqrt{T}\int_0^{M\Delta f} e^{j2\pi f\tau}\frac{1}{T}\sum_m \delta(f-\nu-m\Delta f)df\\
&= \frac{1}{\sqrt{T}}\sum_m\left(\int_0^{M\Delta f} e^{j2\pi f\tau}\delta(f-\nu-m\Delta f)df\right),
\end{aligned}\tag{5.42}$$

where

$$\begin{aligned}
&\int_0^{M\Delta f} e^{j2\pi f\tau}\delta(f-\nu-m\Delta f)\\
&= \begin{cases} e^{j2\pi(\nu+m\Delta f)\tau} & \text{if } 0\le\nu+m\Delta f < M\Delta f,\\ 0 & \text{otherwise.}\end{cases}
\end{aligned}\tag{5.43}$$

Since $0 \le \nu + m\Delta f \le M\Delta f$, the summation over m only needs to be done for $-\lfloor \frac{\nu}{\Delta f} \rfloor \le m < M - \lfloor \frac{\nu}{\Delta f} \rfloor$, where $\lfloor x \rfloor$ denotes the greatest integer less than or equal to $x \in \mathbb{R}$. We have

$$\begin{aligned} Z_T\big[h_{\mathrm{B}}(t)\big](\tau,\nu) &= \frac{1}{\sqrt{T}} \sum_{m=-\lfloor \frac{\nu}{\Delta f} \rfloor}^{M-1-\lfloor \frac{\nu}{\Delta f} \rfloor} e^{j2\pi(\nu+m\Delta f)\tau} \\ &= \frac{1}{\sqrt{T}} e^{-j2\pi \lfloor \frac{\nu}{\Delta f} \rfloor \Delta f \tau} \sum_{m=0}^{M-1} e^{j2\pi(\nu+m\Delta f)\tau} \\ &= \frac{1}{\sqrt{T}} e^{j2\pi\nu\tau} e^{-j2\pi \lfloor \frac{\nu}{\Delta f} \rfloor \Delta f \tau} \sum_{m=0}^{M-1} e^{j2\pi m\Delta f \tau} \\ &= \frac{1}{\sqrt{T}} e^{j2\pi\nu\tau} e^{-j2\pi \lfloor \frac{\nu}{\Delta f} \rfloor \Delta f \tau} e^{j\pi(M-1)\Delta f \tau} \frac{\sin(\pi M \Delta f \tau)}{\sin(\pi \Delta f \tau)}. \end{aligned} \tag{5.44}$$

The ZT restricted to the fundamental region can be written as

$$\bar{Z}_T\big[h_{\mathrm{B}}(t)\big](\tau,\nu) = \frac{1}{\sqrt{T}} e^{j2\pi\nu\tau} e^{j\pi(M-1)\Delta f \tau} \frac{\sin(\pi M \Delta f \tau)}{\sin(\pi \Delta f \tau)}, \tag{5.45}$$

where $(\tau,\nu) \in [0,T] \times [0,1/T]$. □

Proof of (5.41). Here we derive $Z_{\mathrm{T}}[h_{\mathrm{T}}(t)](\tau,\nu)$ starting from

$$Z_T\big[h_{\mathrm{T}}(t)\big](\tau,\nu) = \sqrt{T} \sum_n h_{\mathrm{T}}(\tau + nT) e^{-j2\pi nT\nu}. \tag{5.46}$$

From (5.37), $h_{\mathrm{T}}(\tau + nT)$ is nonzero only for $0 \le \tau + nT < NT$. The infinite summation over n is then limited to $-\lfloor \frac{\tau}{T} \rfloor \le n < N - \lfloor \frac{\tau}{T} \rfloor$. We have

$$\begin{aligned} Z_T\big[h_{\mathrm{T}}(t)\big](\tau,\nu) &= \sqrt{T} \sum_{n=-\lfloor \frac{\tau}{T} \rfloor}^{N-1-\lfloor \frac{\tau}{T} \rfloor} e^{-j2\pi nT\nu} \\ &= \sqrt{T} e^{-j2\pi \lfloor \frac{\tau}{T} \rfloor T} \sum_{n=0}^{N-1} e^{-j2\pi nT\nu} \\ &= \sqrt{T} e^{-j2\pi \lfloor \frac{\tau}{T} \rfloor T\nu} e^{-j\pi\nu(N-1)T} \frac{\sin(\pi\nu NT)}{\sin(\pi\nu T)}. \end{aligned} \tag{5.47}$$

The ZT restricted to the fundamental region can be written as

$$\bar{Z}_T\big[h_{\mathrm{T}}(t)\big](\tau,\nu) = \sqrt{T} e^{-j\pi\nu(N-1)T} \frac{\sin(\pi\nu NT)}{\sin(\pi\nu T)}, \tag{5.48}$$

(a) $N = 16, M = 16$

(b) $N = 128, M = 128$

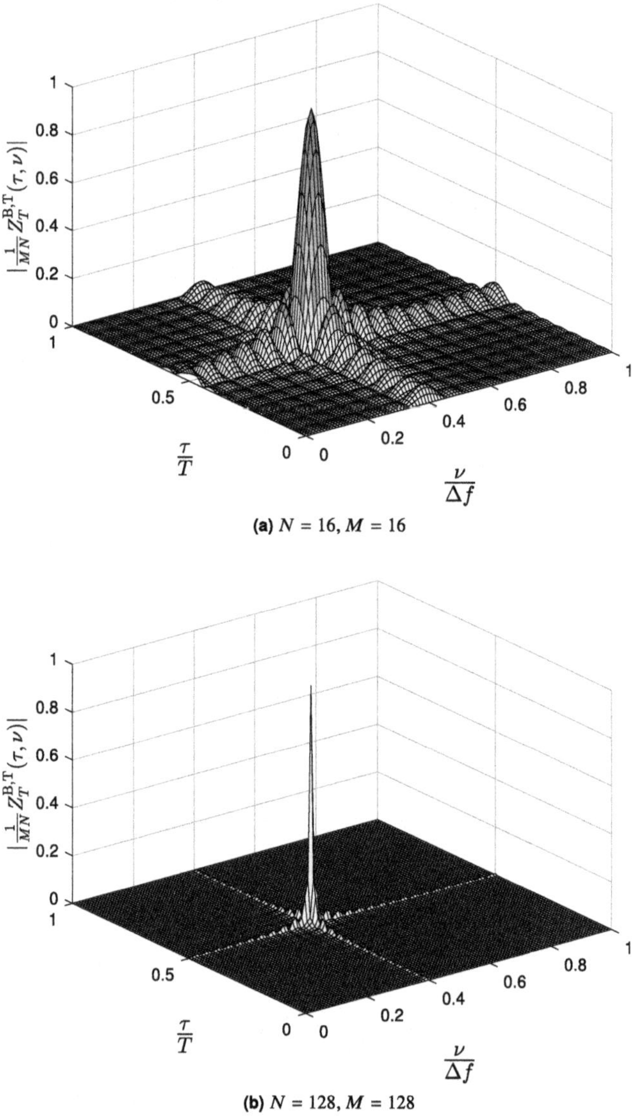

FIGURE 5.5 $|\frac{1}{MN} Z_T^{\text{B,T}}(\tau, \nu)|$ vs. normalized delay ($\frac{\tau}{T}$) and normalized Doppler shift ($\frac{\nu}{\Delta f}$) for $\tau_0 = 0.5T$ and $\nu_0 = 0.4\Delta f$.

where $(\tau, \nu) \in [0, T] \times [0, 1/T]$.

In Fig. 5.5, we plot the magnitude of the BT delay-Doppler basis functions $|\frac{1}{MN} Z_T^{\text{B,T}}(\tau, \nu)|$ for different N and M. It can be noted, due to time and bandwidth limitation, that the BT basis functions are no longer an impulse in the delay-Doppler domain. However, most of the energy of $Z_T[\Phi_{\tau_0, \nu_0}(t)]$

is concentrated around (τ_0, ν_0) in an interval of width $\frac{2}{M\Delta f}$ and $\frac{2}{NT}$ along the delay and Doppler dimensions, respectively. As we increase N and M (see Fig. 5.5b), the BT delay-Doppler functions approach the infinite time and bandwidth case (delta pulses at (τ_0, ν_0)). □

5.4.4 Communications using band- and time-limited signals

Consider a transmit signal

$$s(t) = \sum_i a_i \Phi^{\mathrm{B,T}}_{\tau_i,\nu_i}(t),$$

where a finite number of information symbols a_i are multiplexed on the band- and time-limited basis functions $\Phi^{\mathrm{B,T}}_{\tau_i,\nu_i}(t)$ for $0 \leq t < NT$. The Zak transform of such a signal can be written in the delay-Doppler domain as

$$Z_T\big[s(t)\big](\tau, \nu) = \sum_i a_i Z_T^{\mathrm{B,T}}\big[\Phi_{\tau_i,\nu_i}(t)\big](\tau, \nu).$$

The information symbols a_i can be perfectly recovered from $s(t)$ if these basis functions are orthogonal at the sampling points (τ_i, ν_i) in the delay-Doppler plane. However, as shown in Fig. 5.5 the delay-Doppler basis functions are two-dimensional periodic sinc functions that may introduce ISI across information symbols if the locations (τ_i, ν_i) are not chosen appropriately.

It can be noted from (5.40) and (5.41) that these two-dimensional periodic sinc functions have their zeros at integer multiples of $\frac{1}{M\Delta f}$ and $\frac{1}{NT}$ (the delay-Doppler grid Γ in (4.1)) along the delay and Doppler dimensions, respectively. Leveraging on this property, if we restrict τ_i and ν_i to be integer multiples of $\frac{1}{M\Delta f}$ and $\frac{1}{NT}$, then the delay-Doppler domain basis functions sampled at the (l, k) grid points in Γ

$$Z_T^{\mathrm{B,T}}\big[\Phi_{\tau_i,\nu_i}(t)\big]\left(\frac{l}{M\Delta f}, \frac{k}{NT}\right) = \delta_{l,l_i}\delta_{k,k_i}, \tag{5.49}$$

where $\tau_i = \frac{l_i}{M\Delta f}$ and $\nu_i = \frac{k_i}{NT}$ for $l_i, k_i \in \mathbb{Z}$.

Fig. 5.6 shows the contour plot of the delay-Doppler domain when an information symbol a_i is placed at $\tau_i = 5/M\Delta f$ and $\nu_i = 4/NT$. The red (gray in print version) dots indicate the integer sampling points in the delay-Doppler plane, where other information symbols can be multiplexed, since the delay-Doppler Zak basis functions remain perfectly orthogonal.

Now consider the case where we transmit a single information symbol a_0 over a multipath channel using a BT delay-Doppler domain basis

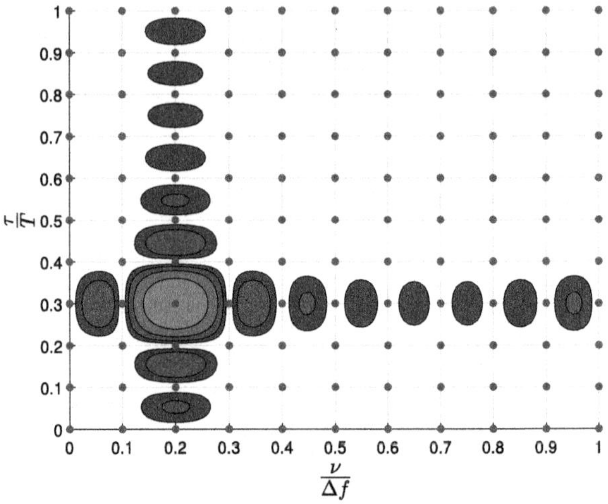

FIGURE 5.6 Contour plot of $|\bar{Z}_T^{\mathrm{B,T}}[s(t)](\tau,\nu)|$ vs. normalized delay ($\frac{\tau}{T}$) and normalized Doppler shift ($\frac{\nu}{\Delta f}$) for $M=10$, $N=10$ for one information symbol placed at $\tau_0 = 3/M\Delta f$ and $\nu_0 = 2/NT$.

function, i.e., $s(t) = a_0 \Phi^{\mathrm{B,T}}_{\tau_0,\nu_0}(t)$. Then

$$r(t) = \sum_{p=1}^{P} g_p e^{j2\pi\nu_p(t-\tau_p)} a_0 \Phi^{\mathrm{B,T}}_{\tau_0,\nu_0}(t-\tau_p),$$

where g_p is the path gain and τ_p and ν_p are the delay and the Doppler shift associated with the p-th path. We then have the Zak transform of the received signal as

$$\boxed{Z_T\big[r(t)\big](\tau,\nu) = a_0 \sum_{p=1}^{P} g_p e^{j2\pi\nu_p(\tau-\tau_p)} Z_T^{\mathrm{B,T}}\big[\Phi_{\tau_0,\nu_0}(t)\big](\tau-\tau_p,\nu-\nu_p).}$$

Fig. 5.7 shows the contour plot of the received signal in the delay-Doppler domain if a single information symbol a_0 is transmitted at $(\tau_0,\nu_0) = (3/M\Delta f, 2/NT)$ through a channel with $P=2$ paths with $(\tau_1,\nu_1) = (0/M\Delta f, 0/NT)$ and $(\tau_2,\nu_2) = (3/M\Delta f, 3/NT)$. For ease of visualization, we consider $g_1 = g_2 = 1$.

Following (5.49), it can be observed that if $Z_T[r(t)](\tau,\nu)$ is sampled at integer multiples of $1/M\Delta f$ and $1/NT$ (denoted by the red (gray in print version) dots), then out of the NM sampling points, there are only *two* points with nonzero values at the receiver, thereby limiting the intersymbol interference to at most P other information symbols. Note that

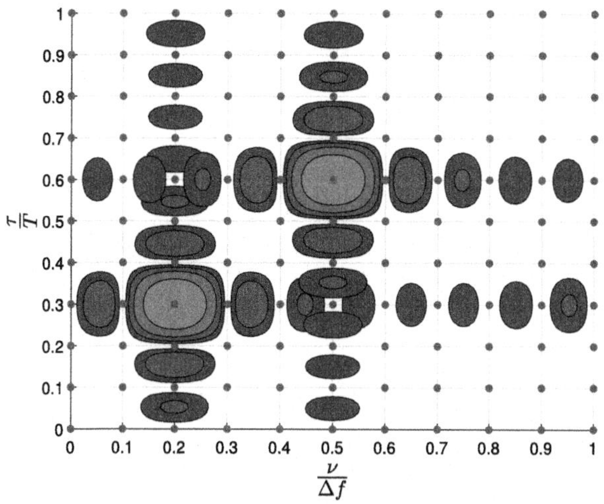

FIGURE 5.7 Contour plot of $|Z_T[r(t)](\tau, \nu)|$ vs. normalized delay ($\frac{\tau}{T}$) and normalized Doppler shift ($\frac{\nu}{\Delta f}$) for $M = 10$, $N = 10$, $\tau_0 = 3/M\Delta f$, and $\nu_0 = 2/NT$ and channel parameters $(\tau_1, \nu_1) = (0/M\Delta f, 0/NT)$ and $(\tau_2, \nu_2) = (3/M\Delta f, 3/NT)$.

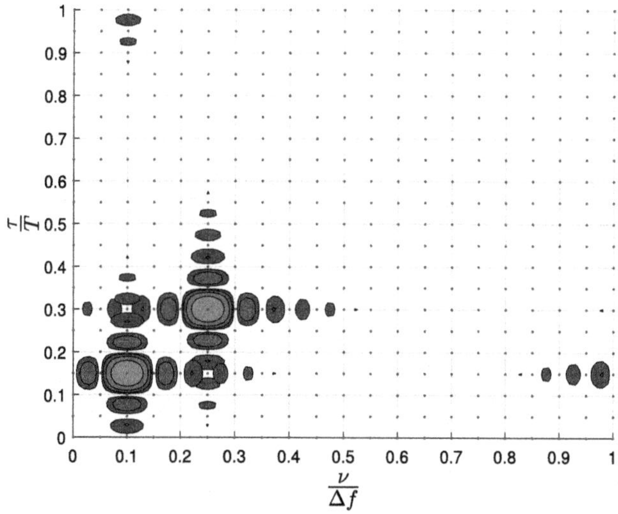

FIGURE 5.8 Contour plot of $|Z_T[r(t)](\tau, \nu)|$ vs. normalized delay ($\frac{\tau}{T}$) and normalized Doppler shift ($\frac{\nu}{\Delta f}$) for $M = 20$, $N = 20$, $\tau_0 = 3/M\Delta f$, and $\nu_0 = 2/NT$ and channel parameters $(\tau_1, \nu_1) = (0/M\Delta f, 0/NT)$ and $(\tau_2, \nu_2) = (3/M\Delta f, 3/NT)$.

sampling at noninteger multiples may lead to significant ISI. However, the number of significant terms in the ISI can be still limited (relative to MN) by increasing N and M as shown in Fig. 5.8, where the time duration and bandwidth of $s(t)$ is doubled compared to the case in Fig. 5.7.

In summary, the sparsity of the channel in the continuous delay-Doppler domain can be retained by choosing the appropriate basis functions at the receiver as well as the duration and bandwidth of the frame. Given a sparse channel with P paths with integer delay and Doppler shifts, the ISI in the delay-Doppler domain can be limited to P other symbols, if the delay and Doppler dimensions are sampled at integer multiples of $1/M\Delta f$ and $1/NT$. This uniform sampling of the continuous delay-Doppler domain leads us to a finite version of the Zak transform known as the discrete Zak transform (DZT). The Zak analysis of delay-Doppler communications can be further simplified by using the DZT.

5.5 The discrete Zak transform

Let us now consider a discrete-time signal $\bar{x}[q], q = 0, \dots NM - 1$ obtained by sampling $x(t)$ with sampling period $T_s = T/M$ over a window of size NT. We define the *discrete Zak transform* of $\bar{x}[q]$ *over the infinite grid* $[l, k] \in \mathbb{Z}^2$ as[1]

$$\begin{aligned} Z_{\bar{x}}[l,k] &= \sum_n \sum_m \bar{Z}_{\bar{x}}[l+nM, k+mN] e^{j\frac{2\pi}{N}n(k+mN)} \\ &= \sum_n \sum_m \bar{Z}_{\bar{x}}[l+nM, k+mN] e^{j\frac{2\pi}{N}nk}, \end{aligned} \tag{5.50}$$

and

$$\bar{Z}_{\bar{x}}[l,k] = \frac{1}{\sqrt{N}} \sum_{n'=0}^{N-1} \bar{x}[l+n'M] e^{-j\frac{2\pi}{N}n'k}, \quad \begin{array}{l} 0 \le l \le M-1, \\ 0 \le k \le N-1 \end{array} \tag{5.51}$$

is the *DZT defined only in the fundamental region* $\mathcal{R} = [0, M-1] \times [0, N-1]$ and zero elsewhere.

We note that for any given delay (or time shift) $0 \le l < M - 1$, the sum in (5.51) is the N-point DFT of the discrete-time signal $\bar{x}_l[n] = \bar{x}[l + nM]$ obtained by decimating $\bar{x}[q]$ M times, with an offset of l samples. The $\bar{Z}_{\bar{x}}[l,k]$ can be conveniently represented by an $M \times N$ matrix $\mathbf{Z_x}$, where the vector $\mathbf{x}$ has NM components $\bar{x}[q]$.

Similarly to the Fourier transform, the DZT is the result of a projection of the time domain signal $x[q]$ onto the time domain DZT basis functions parametrized by l and k:

$$Z_{\bar{x}}[l, k+mN] = \langle x[q], \Phi_{l,k}[q] \rangle = \sum_{q=-\infty}^{+\infty} x[q] \Phi_{l,k}[q]^* \tag{5.52}$$

[1]In this chapter, we follow the conventional notations used in the discrete Zak transform literature with l and k for delay and Doppler indices, different from m and n used in Chapter 4.

for any integer n (due to the periodicity in k), where

$$\Phi_{l,k}[q] = \sum_{n=-\infty}^{+\infty} \delta[q - l - nM] e^{j\frac{2\pi}{N}nk}. \tag{5.53}$$

These Zak basis functions are pulse trains shifted by $\tau = l/M\Delta f$ and modulated by a complex sine wave at frequency $\nu = k/NT$.

5.5.1 The inverse discrete Zak transform

The inverse discrete Zak transform (IDZT) is given by

$$\begin{aligned}\bar{x}[q] = \bar{x}[l + nM] = \mathrm{IDZT}_M\{Z_{\bar{x}}[l,k]\} &= \frac{1}{\sqrt{N}} \sum_{k=0}^{N-1} Z_{\bar{x}}[l,k] e^{j\frac{2\pi}{N}nk} \\ &= \frac{1}{\sqrt{N}} \sum_{k=0}^{N-1} \bar{Z}_{\bar{x}}[l,k] e^{j\frac{2\pi}{N}nk}, \qquad \begin{array}{l} 0 \le l \le M-1, \\ 0 \le n \le N-1. \end{array}\end{aligned} \tag{5.54}$$

We can write the inverse Zak transform of the vector $\mathbf{x}$ in a compact form as

$$\mathbf{x} = \mathrm{vec}\big[\mathrm{IDFT}_N(\mathbf{Z}_\mathbf{x})\big],$$

where the N-point IDFT is applied to each row of $\mathbf{Z}_\mathbf{x}$ followed by the row-column deinterleaver as illustrated in Fig. 4.6 for an OTFS modulator transforming the delay-Doppler information symbols in $\mathbf{X}$ to the time domain vector $\mathbf{s}$.

5.5.2 Properties of the DZT

All properties of the Zak transform apply for the DZT if we consider $x[q]$ as the periodic extension of $\bar{x}[q]$ and use the extended DZT in (5.50). We list below the most relevant ones.

1. Periodic by N along the Doppler axis and quasiperiodic by M along delay axis:

$$Z_x[l,k] = Z_x\big[[l]_M, [k]_N\big] e^{j\frac{2\pi}{N}k\lfloor\frac{l}{M}\rfloor}, \quad [l,k] \in \mathbb{Z}^2. \tag{5.55}$$

2. Time shift by $q_0 = l_0 + n_0 M$ with $0 \le l_0 < M-1$:

$$\mathrm{DZT}_M\{x[q + q_0]\}[l,k] = e^{j\frac{2\pi}{N}n_0k} Z_x[l + l_0, k]. \tag{5.56}$$

3. Frequency shift by k_0 (modulation at frequency k_0):

$$\mathrm{DZT}_M\{x[q]e^{j\frac{2\pi}{NM}k_0q}\}[l,k] = e^{j\frac{2\pi}{NM}k_0l} Z_x[l, k - k_0]. \tag{5.57}$$

4. Product of $w[q]$ and $x[q]$ both periodic with period NM:

$$\mathrm{DZT}_M\{w[q]\cdot x[q]\}[l,k]=\frac{1}{\sqrt{N}}\sum_{k'=0}^{N-1} Z_w[l,k']\,Z_x[l,k-k'], \tag{5.58}$$

i.e., the DZT of the product in time is convolution of the DZTs along the Doppler axis.

5. Convolution of $w[q]$ and $x[q]$ both periodic with period NM:

$$\mathrm{DZT}_M\{w[q]*x[q]\}[l,k]=\sqrt{N}\sum_{l'=0}^{M-1} Z_w[l',k]\,Z_x[l-l',k], \tag{5.59}$$

i.e., the DZT of the convolution in time is convolution of the DZTs along the delay axis.

Note that these properties involve the infinite extension of the Zak transform Z_x and cannot be reduced to $\bar{Z}_x$, due to the quasiperiodicity in the delay axis.

5.6 DZT in delay-Doppler communications

Consider the discrete-time transmitted signal frame $s[q]$ with NM samples generated as the IDZT of $Z_s[l,k]$ as illustrated in Fig. 4.6. The NM information symbols in the frame are placed in the delay-Doppler domain as the values of $\bar{Z}_s[l,k]$ for $[l,k]\in\mathcal{R}$.

We assume a time-varying multipath channel that has P paths in the delay-Doppler domain, each having the propagation gain g_i, the integer delay and Doppler taps $l_i=\tau_i/M\Delta f$ and $k_i=\nu_i/NT$, for $i=1,\ldots,P$. For simplicity, we omit the AWGN term in the following analysis.

5.6.1 Receiver sampling

The discrete-time received signal obtained by sampling the continuous-time received waveform at $f_s=M\Delta f$ [Hz] is obtained as (see Chapter 2)

$$\boxed{r[q]=\sum_{i=1}^{P} g_i e^{\frac{j2\pi}{NM}k_i(q-l_i)}s[q-l_i],} \tag{5.60}$$

where we assume $s[q]$ is periodic in time with period NM and is modulated using IDZT of the fundamental region $\bar{Z}_s[l,k]$ with $[l,k]\in\mathcal{R}$. Using

(5.51), we obtain

$$\begin{aligned}\bar{Z}_r[l,k] &= \frac{1}{\sqrt{N}}\sum_{n=0}^{N-1}\sum_{i=1}^{P} g_i e^{\frac{j2\pi}{NM}k_i(l+nM-l_i)} s[l+nM-l_i] e^{-\frac{j2\pi}{N}nk} \\ &= \sum_{i=1}^{P} g_i e^{\frac{j2\pi}{NM}k_i(l-l_i)} \frac{1}{\sqrt{N}}\sum_{n=0}^{N-1} s[l-l_i+nM] e^{\frac{-j2\pi}{N}n(k-k_i)} \\ &= \sum_{i=1}^{P} g_i e^{\frac{j2\pi}{NM}k_i(l-l_i)} Z_s[l-l_i,k-k_i]. \end{aligned} \tag{5.61}$$

This does not yet represent the delay-Doppler input–output relation relating NM transmitted information symbols in $\bar{Z}_s[l,k]$ to the NM received samples in $\bar{Z}_r[l,k]$ inside the fundamental region.

From (5.61) we note that for $l-l_i \notin [0, M-1]$ and $k-k_i \notin [0, N-1]$, $Z_s[l-l_i,k-k_i]$ falls outside of the fundamental rectangle $\mathcal{R}$ for $[l,k] \in \mathcal{R}$. Thanks to the periodic and quasiperiodic property of the DZT transform given in (5.55) we have

$$Z_s[l-l_i,k-k_i] = \bar{Z}_s\big[[l-l_i]_M,[k-k_i]_N\big]e^{j\frac{2\pi}{N}(k-k_i)\lfloor\frac{l-l_i}{M}\rfloor}, \tag{5.62}$$

where $\lfloor\frac{l-l_i}{M}\rfloor$ only takes the value 0 for $l \geq l_i$ and -1 for $l < l_i$, since $0 \leq l$, $l_i \leq M-1$. Then

$$Z_s[l-l_i,k-k_i] = \begin{cases} \bar{Z}_s[[l-l_i]_M,[k-k_i]_N]e^{-j\frac{2\pi}{N}(k-k_i)} & l < l_i, \\ \bar{Z}_s[[l-l_i]_M,[k-k_i]_N] & l \geq l_i. \end{cases} \tag{5.63}$$

Replacing (5.63) in (5.61) yields the delay-Doppler input–output relation for the periodic signals $s[q]$ and $r[q]$ with period NM:

$$\bar{Z}_r[l,k] = \begin{cases} \sum_{i=1}^{P} g_i e^{\frac{j2\pi}{NM}k_i[l-l_i]_M} e^{-j\frac{2\pi}{N}k}\bar{Z}_s[[l-l_i]_M,[k-k_i]_N], \\ \sum_{i=1}^{P} g_i e^{\frac{j2\pi}{NM}k_i[l-l_i]_M}\bar{Z}_s[[l-l_i]_M,[k-k_i]_N]. \end{cases} \tag{5.64}$$

5.6.2 Time-windowing at RX and TX

The time domain signal generated through the IDZT of $Z_s[l,k]$ for $[l,k] \in \mathbb{Z}$ is periodic with period NM. However, in practical applications, a window $w_{\text{tx}}[q]$ is applied at the transmitter to generate a signal of finite duration NT [s] or NM samples. Extra samples can be appended to the time domain waveform to combat the effects of multipath fading in the different variants of OTFS discussed in Chapter 4.

Similarly, at the receiver a window $w_{\text{rx}}[q]$ is applied to acquire the NM samples needed to reconstruct $\bar{Z}_r[l,k]$ and subsequently estimate $\bar{Z}_s[l,k]$

(through any detection method) for $[l,k] \in \mathcal{R}$. The received time domain signal can be written with transmit and receive windowing as

$$\boxed{r[q] = w_{\text{rx}}[q] \sum_{i=1}^{P} g_i e^{j\frac{2\pi}{NM}k_i(q-l_i)} w_{\text{tx}}[q-l_i] s[q-l_i],} \tag{5.65}$$

where $w_{\text{tx}}[q] = 1$ for $0 \leq q < NM$ and *zero* otherwise, and $l_i \leq l_{\max}$, where $l_{\max}$ is the maximum channel delay tap. The DZT of the received signal can be written as

$$\begin{aligned} \bar{Z}_r[l,k] =& \frac{1}{\sqrt{N}} \sum_{n=0}^{N-1} w_{\text{rx}}[l+nM] \times \\ & \sum_{i=1}^{P} g_i e^{j\frac{2\pi}{NM}k_i(l+nM-l_i)} w_{\text{tx}}[l+nM-l_i] s[l+nM-l_i] e^{-j\frac{2\pi}{N}nk} \\ =& \sum_{i=1}^{P} g_i e^{j\frac{2\pi}{NM}k_i(l-l_i)} \times \\ & \underbrace{\frac{1}{\sqrt{N}} \sum_{n=0}^{N-1} \underbrace{w_{\text{rx}}[l+nM] w_{\text{tx}}[l+nM-l_i]}_{w_{\text{eff}}[l+nM]} s[l+nM-l_i] e^{-j\frac{2\pi}{N}n(k-k_i)}}_{(a)}, \end{aligned} \tag{5.66}$$

where the term $w_{\text{eff}}[l+nM]$ is acting as an *effective time domain window* on the propagation delay l_i of each path, given by

$$\boxed{w_{\text{eff}}[l+nM] = w_{\text{rx}}[l+nM] w_{\text{tx}}[l+nM-l_i]} \tag{5.67}$$

for $l = 0 \ldots M-1$ and $n = 0 \ldots N-1$. The term (a) in (5.66) is the Zak transform of the product of the effective time domain window and the channel impaired transmitted signal, which can be written as

$$\begin{aligned} \frac{1}{\sqrt{N}} & \sum_{n=0}^{N-1} \left(w_{\text{eff}}[l+nM] \cdot s[l+nM-l_i] \right) e^{-j\frac{2\pi}{N}n(k-k_i)} \\ & = \text{DZT}\left\{ w_{\text{eff}}[q] \cdot s[q-l_i] \right\}[l, k-k_i] \\ & = Z_{w_{\text{eff}} \cdot s}[l-l_i, k-k_i]. \end{aligned} \tag{5.68}$$

Using the property in (5.58), the Zak transform of the product of time domain signals in (5.68) results in the convolution along the Doppler domain of the respective Zak transforms as

$$Z_{w_{\text{eff}}\cdot s}[l-l_i,k-k_i]=\frac{1}{\sqrt{N}}\sum_{k'=0}^{N-1}Z_{w_{\text{eff}}}\left[l-l_i,k-k_i-k'\right]Z_s\left[l-l_i,k'\right]. \tag{5.69}$$

The input–output relation in (5.66) after including the effect of windowing becomes

$$\bar{Z}_r[l,k]=\sum_{i=1}^{P}g_i e^{j\frac{2\pi}{NM}k_i(l-l_i)}Z_{w_{\text{eff}}\cdot s}[l-l_i,k-k_i]. \tag{5.70}$$

The input–output relation in (5.70) can be easily modified for any arbitrary window by substituting the DZT of the effective window function for $Z_{w_{\text{eff}}}$ in (5.69).

5.6.3 RCP-OTFS with rectangular Tx and Rx window

Reduced cyclic prefix OTFS (RCP-OTFS) is the OTFS variant (see Chapter 4) where a CP of sufficient length $L_{\text{CP}} \geq l_{\max}$ is added to an OTFS frame. Here we set $L_{\text{CP}} = l_{\max} < M$. Among OTFS variants, RCP-OTFS is the one most closely related to the DZT.

By using the transmitter and receiver windows of length $(NM + L_{\text{CP}} = NM + l_{\max})$ and (NM), respectively, a channel with maximum delay spread $l_{\max}$ interacts with the transmitted signal as if the periodic and infinite-length signal $s[q]$ was transmitted, even though in reality the finite-length $\bar{s}[q]$ was transmitted. The transmitter window for RCP-OTFS operates on the periodic signal $\bar{s}[q]$ generated from $\bar{Z}_s$ as

$$s[q]=w_{\text{tx}}[q]\bar{s}[q],$$

where

$$w_{\text{tx}}[q]=1 \ \text{ for } \ -L_{\text{CP}} \leq q < NM,$$

which restricts the periodic signal $\bar{s}[q]$ to the sample indices $[-L_{\text{CP}}, NM-1]$. At the receiver side NM samples are acquired using the window

$$w_{\text{rx}}[q]=1 \ \text{ for } \ 0 \leq q < NM.$$

As shown in Fig. 5.9(a), the transmitter window (black) is longer than the receiver window (red (mid gray in print version)) and is equivalent to adding a CP of length $L_{\text{CP}} = l_{\max}$ at the transmitter. As illustrated in the figure, the effective window function (green (light gray in print version))

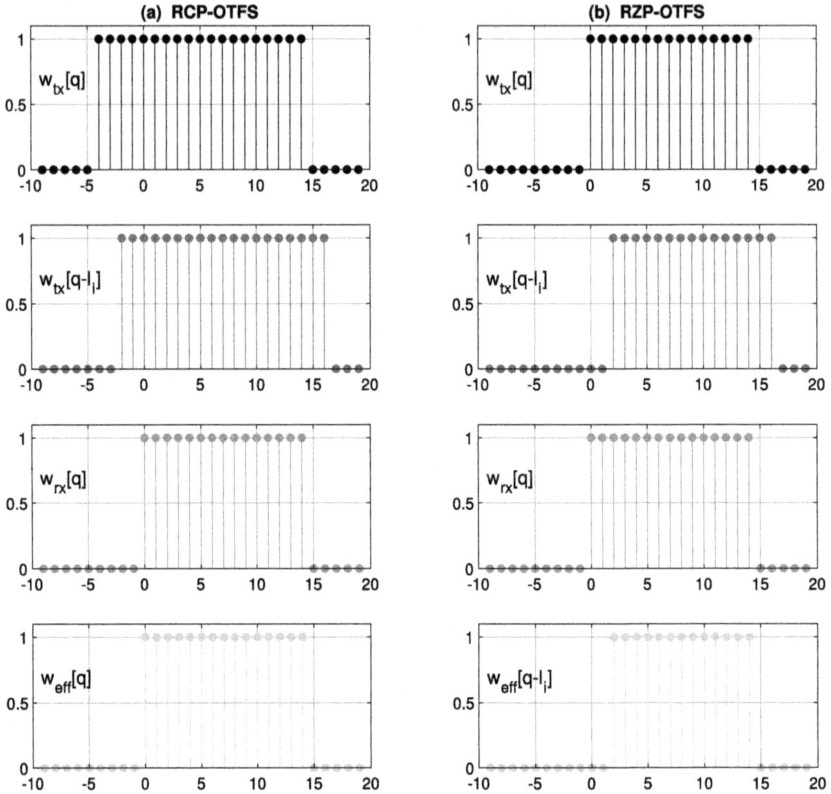

FIGURE 5.9 The rectangular transmit $w_{\text{tx}}[q]$ (black), delayed transmit $w_{\text{tx}}[q-l_i]$ (blue (dark gray in print version)), receiver $w_{\text{rx}}[q]$ (red (mid gray in print version)), and effective window functions $w_{\text{tx}}[q]$ (green (light gray in print version)) for RCP-OTFS (left) and $w_{\text{tx}}[q-l_i]$ (green (light gray in print version)) for RZP-OTFS (right) for $M=5$, $N=3$, $L_{\text{CP}}=L_{\text{ZP}}=4$, and $l_i=2$.

is the product of the receiver (red (mid gray in print version)) and delayed transmit (blue (dark gray in print version)) window functions, given as

$$\underbrace{w_{\text{eff}}[q]}_{\text{green}} = w_{\text{eff}}[l+nM] = \underbrace{w_{\text{rx}}[l+nM]}_{\text{red}}\,\underbrace{w_{\text{tx}}[l+nM-l_i]}_{\text{blue}} = 1 \tag{5.71}$$

for $l_i \le l_{\text{max}}$, where $l=0,\ldots,M-1$ and $n=0,\ldots,N-1$, and zero elsewhere. Taking DZT of (5.71) yields

$$Z_{w_{\text{eff}}}[l-l_i,k] = \frac{1}{\sqrt{N}}\sum_{n=0}^{N-1} w_{\text{eff}}[l+nM-l_i]e^{-j\frac{2\pi}{N}kn} = \frac{1}{\sqrt{N}}\zeta_N(-k), \tag{5.72}$$

where $\zeta_N(\cdot)$ is the periodic sinc function defined in (4.79). Substituting this into (5.69) yields the DZT of the time domain product of the transmit signal with the effective window as

$$Z_{w_{\text{eff}}\cdot s}[l-l_i,k-k_i]=\frac{1}{N}\sum_{k'=0}^{N-1}\zeta_N\big(k'-(k-k_i)\big)Z_s\big[l-l_i,k'\big]. \tag{5.73}$$

For integer k_i, $\zeta_N(x)=N$ for $x=0$ and *zero* otherwise. The above equation becomes

$$Z_{w_{\text{eff}}\cdot s}[l-l_i,k-k_i]=Z_s[l-l_i,k-k_i]. \tag{5.74}$$

Substituting (5.74) in (5.70) and applying the property in (5.63), the delay-Doppler input–output relations can be written in the general form as

$$\boxed{\bar{Z}_r[l,k]=\sum_{i=1}^{P}g_i\phi_i[l,k]\bar{Z}_s\big[[l-l_i]_M,[k-k_i]_N\big],} \tag{5.75}$$

where the following phase rotations are given in Table 4.3 as

$$\phi_i[l,k]=\begin{cases} e^{j\frac{2\pi}{NM}k_i[l-l_i]}, & \text{if } l\geq l_i,\\ e^{j\frac{2\pi}{NM}k_i[l-l_i]_M}e^{-j\frac{2\pi}{N}k}, & \text{if } l<l_i.\end{cases} \tag{5.76}$$

The extra phase for the case when $l<l_i$ is due to the Zak property for translation out of the fundamental region in (5.62). This delay-Doppler input–output relation coincides with (5.64) for NM-periodic $s[q]$ and $r[q]$, and with the one derived for RCP-OTFS given in (4.120) and (4.121), where the index m is replaced by l.

5.6.4 RZP-OTFS with rectangular Tx and Rx window

Reduced zero padding OTFS (RZP-OTFS) is another OTFS variant presented in Chapter 4, which can be formed by adding a ZP of sufficient length $L_{\text{ZP}}\geq l_{\max}$ to an OTFS frame. Here, we set $L_{\text{ZP}}=l_{\max}$. In RZP-OTFS, the periodic assumption of the transmit waveform is discarded and the channel interacts with $s[q]$ rather than $\bar{s}[q]$. Let

$$w_{\text{tx}}[q]=1 \text{ for } 0\leq q<NM,\quad w_{\text{rx}}[q]=1 \text{ for } 0\leq q<NM,$$

which are shown in black and red (mid gray in print version) in Fig. 5.9(b). Unlike RCP-OTFS, both windows in RZP-OTFS have the same size. The effective window function is given as

$$w_{\text{eff}}[q-l_i]=w_{\text{eff}}[l+nM-l_i]=\begin{cases}1, & \text{for } l_i\leq q<NM,\\ 0, & \text{otherwise}\end{cases} \tag{5.77}$$

for $l = 0, \ldots, M-1$ and $n = 0, \ldots, N-1$. Fig. 5.9(b) shows the effective window in green (light gray in print version). The maximum value of l_i is assumed to be less than the channel delay spread $l_{\max}$. The DZT of (5.77) is given as

$$Z_{w_{\text{eff}}}[l-l_i,k] = \frac{1}{\sqrt{N}} \sum_{n=0}^{N-1} w_{\text{eff}}[l+nM-l_i] e^{-j\frac{2\pi}{N}kn}$$
$$= \begin{cases} \frac{1}{\sqrt{N}} \zeta_N(-k), & \text{for } l \geq l_i, \\ \frac{1}{\sqrt{N}} (\zeta_N(-k)-1), & \text{for } l < l_i. \end{cases} \tag{5.78}$$

Substituting it in (5.69) yields

$$Z_{w_{\text{eff}} \cdot s}[l-l_i, k-k_i] = \frac{1}{\sqrt{N}} \sum_{k'=0}^{N-1} Z_{w_{\text{eff}}}\big[l-l_i, k-k_i-k'\big] Z_s\big[l-l_i, k'\big], \tag{5.79}$$

where for integer values of k_i, we have

$$\frac{1}{\sqrt{N}} Z_{w_{\text{eff}}}\big[l-l_i, (k-k_i)-k'\big] = \begin{cases} 1, & \text{if } l \geq l_i, k' = k-k_i, \\ \frac{N-1}{N}, & \text{if } l < l_i, k' = k-k_i, \\ -\frac{1}{N}, & \text{if } l < l_i, k' \neq k-k_i, \\ 0, & \text{otherwise.} \end{cases} \tag{5.80}$$

Substituting (5.80) in (5.70) and applying the property in (5.63), the RZP-OTFS input–output relation for integer k_i reduces to

$$\boxed{\bar{Z}_r[l,k] = \sum_{i=1}^{P} g_i e^{j\frac{2\pi}{NM}k_i(l-l_i)} \sum_{k'=0}^{N-1} \beta_i\big[l,(k-k_i)-k'\big] \bar{Z}_s\big[[l-l_i]_M, k'\big],} \tag{5.81}$$

where

$$\beta_i\big[l,(k-k_i)-k'\big] = \begin{cases} 1, & \text{if } l \geq l_i, k' = [k-k_i]_N, \\ \frac{N-1}{N} e^{-j2\pi([k-k_i]_N)/N}, & \text{if } l < l_i, k' = [k-k_i]_N, \\ -\frac{1}{N} e^{-j2\pi k'/N}, & \text{if } l < l_i, k' \neq [k-k_i]_N, \\ 0, & \text{otherwise.} \end{cases} \tag{5.82}$$

For very large values of N, the third term in (5.82) for $k' \neq [k-k_i]_N$ may be ignored to approximate (5.81) as

$$\boxed{\bar{Z}_r[l,k] \approx \sum_{i=1}^{P} g_i \phi_i[l,k] \bar{Z}_s\big[[l-l_i]_M, [k-k_i]_N\big],} \tag{5.83}$$

where the following phase rotations are given in Table 4.3 as

$$\phi_i[l,k] = \begin{cases} e^{j\frac{2\pi}{NM}k_i[l-l_i]}, & \text{if } l \geq l_i, \\ \frac{N-1}{N} e^{j\frac{2\pi}{NM}k_i[l-l_i]} e^{-j\frac{2\pi}{N}[k-k_i]_N}, & \text{if } l < l_i. \end{cases} \tag{5.84}$$

Remark. The Zak analysis for rectangular waveform can be extended for different pulse shaping waveforms by modifying the window function $w_{\text{eff}}[q]$. The only additional task is to find the DZT of the corresponding pulse shaping waveform. The derivation of RCP/RZP-OTFS presented in this chapter can easily be extended to the CP/ZP-OTFS case by simply choosing the appropriate time domain input–output relation for the corresponding OTFS variant as given in Chapter 4.

5.7 Bibliographical notes

The Zak transform was named after the inventor J. Zak [1]. The readers can refer to [2,3] for basic definition and properties of the continuous- and discrete-time Zak transform. The OTFS modulation was proposed by Hadani et al. in the 2017 IEEE Wireless Communications and Networking Conference [4], where the authors have pointed out that OTFS is naturally associated with the Zak transform. Later, the precise relation and analysis of OTFS with the Zak transform was presented in [5,6]. The RZP/RCP-OTFS input–output relations adopted in this chapter were introduced in [7] and [8].

References

[1] J. Zak, Finite translation in solid state physics, Physical Review Letters 9 (1967) 1385–1397.

[2] A. Janssen, The Zak transform: a signal transform for sampled time-continuous signals, Philips Journal Research 43 (1) (Jan. 1988) 23–69.

[3] H. Bölcskei, F. Hlawatsch, Discrete Zak transforms, polyphase transforms, and applications, IEEE Transactions on Signal Processing 45 (4) (1997) 851–866, https://doi.org/10.1109/78.564174.

[4] R. Hadani, S. Rakib, M. Tsatsanis, A. Monk, A.J. Goldsmith, A.F. Molisch, R. Calderbank, Orthogonal time frequency space modulation, in: 2017 IEEE Wireless Communications and Networking Conference (WCNC), 2017, pp. 1–6.

[5] S.K. Mohammed, Derivation of OTFS modulation from first principles, IEEE Transactions on Vehicular Technology 70 (8) (2021) 7619–7636, https://doi.org/10.1109/TVT.2021.3069913.

[6] S.K. Mohammed, Time-domain to delay-Doppler domain conversion of OTFS signals in very high mobility scenarios, IEEE Transactions on Vehicular Technology 70 (6) (2021) 6178–6183, https://doi.org/10.1109/TVT.2021.3071942.

[7] P. Raviteja, K.T. Phan, Y. Hong, E. Viterbo, Interference cancellation and iterative detection for orthogonal time frequency space modulation, IEEE Transactions on Wireless Communications 17 (10) (2018) 6501–6515, https://doi.org/10.1109/TWC.2018.2860011.

[8] P. Raviteja, Y. Hong, E. Viterbo, E. Biglieri, Practical pulse-shaping waveforms for reduced-cyclic-prefix OTFS, IEEE Transactions on Vehicular Technology 68 (1) (2019) 957–961, https://doi.org/10.1109/TVT.2018.2878891.

CHAPTER

6

Detection methods

OUTLINE

Delay-Doppler Communications
https://doi.org/10.1016/B978-0-32-385028-5.00014-1

Chapter points

- Overview of OTFS input–output relation.
- Single-tap frequency domain equalization.
- Linear minimum mean-square error detection.
- Message passing detection.
- Maximum-ratio combining detection.
- Iterative rake turbo decoder.

6.1 Overview of OTFS input–output relation

In this section, we recall the delay-Doppler input–output relation for OTFS using rectangular pulse shaping waveforms in Chapter 4. In a single antenna OTFS system, we assume that a frame of duration NT occupies a bandwidth of $M\Delta f$, where $T = 1/\Delta f$ and Δf is the subcarrier spacing. Let $\mathbf{X} \in \mathbb{C}^{M\times N}$ and $\mathbf{Y} \in \mathbb{C}^{M\times N}$ be the transmitted and received information samples matrices in the delay-Doppler domain. After vectorizing $\mathbf{X}^{\mathrm{T}}$ and $\mathbf{Y}^{\mathrm{T}}$, we obtain the corresponding samples vectors $\mathbf{x} \in \mathbb{C}^{NM\times 1}$ and $\mathbf{y} \in \mathbb{C}^{NM\times 1}$, each containing M blocks of N samples, given by

$$\mathbf{x} = \begin{bmatrix} \mathbf{x}_0 \\ \vdots \\ \mathbf{x}_{M-1} \end{bmatrix} = \mathrm{vec}(\mathbf{X}^{\mathrm{T}}),$$
$$\mathbf{y} = \begin{bmatrix} \mathbf{y}_0 \\ \vdots \\ \mathbf{y}_{M-1} \end{bmatrix} = \mathrm{vec}(\mathbf{Y}^{\mathrm{T}}), \tag{6.1}$$

where the subvectors $\mathbf{x}_m,\ \mathbf{y}_m \in \mathbb{C}^{N\times 1}$ for $m = 0, \ldots, M-1$ are along the Doppler axis. The vectorized input–output relation in the delay-Doppler domain can be written as

$$\boxed{\mathbf{y} = \mathbf{H}\cdot\mathbf{x} + \mathbf{z},} \tag{6.2}$$

where $\mathbf{H} \in \mathbb{C}^{NM \times NM}$ is the delay-Doppler channel matrix, given in (4.60) in Chapter 4, and $\mathbf{z} \in \mathbb{C}^{NM \times 1}$ is the AWGN vector with zero mean and σ_w^2 variance.

The vectorized delay-Doppler input–output relation in (6.2) of the noiseless case was shown in Fig. 4.12 with $M = 8$ for three delay paths. It illustrates that $\mathbf{H}$ is a banded matrix with a maximum bandwidth of $N(l_{\max} + 1)$, where $l_{\max}$ denotes the maximum channel delay tap. Further, the matrix $\mathbf{H}$ contains submatrices $\mathbf{K}_{m,l} \in \mathbb{C}^{N \times N}$ representing the linear time-variant channel between the m-th received block and the $[m - l]_M$-th transmitted block for $l \in \mathcal{L}$, where $\mathcal{L}$ is the set of the distinct normalized delay shifts of the channel. Then, each received samples vector $\mathbf{y}_m \in \mathbb{C}^{N \times 1}$, $m = 0, \ldots, M - 1$, was given in (4.63) as

$$\boxed{\mathbf{y}_m = \sum_{l \in \mathcal{L}} \mathbf{K}_{m,l} \cdot \mathbf{x}_{[m-l]_M} + \mathbf{z}_m,} \tag{6.3}$$

where the entries of $\mathbf{K}_{m,l}$ were given in Table 4.2, and for the case of zero padding OTFS (ZP-OTFS), $\mathbf{x}_{[m-l]_M}$ can be reduced to $\mathbf{x}_{m-l}$, since $\mathbf{x}_{m-l} = 0$ for $m < l$.

Here, matrix $\mathbf{H}$ has S nonzero elements in each row, which corresponds to the number of discrete delay-Doppler paths in the delay-Doppler grid. The multipath channel with P paths may result in $S \geq P$, since fractional delay and/or Doppler cause leakage into neighboring grid points for every one of the P paths, when the receiver is sampling, as detailed in Section 4.6.1. Equality holds ($S = P$) when delay and Doppler shifts of the multipaths are exactly integer multiples of delay resolution $1/M\Delta f$ and Doppler resolution $1/NT$ of an OTFS frame, which is not always a practical assumption. Due to the underspread nature of typical wireless channels, we can assume that $S \ll NM$, which implies that $\mathbf{H}$ is *a sparse matrix*.

The sparsity of matrix $\mathbf{H}$ combined with other special properties can be used to design efficient detectors for OTFS, which will be discussed in detail in the rest of the chapter.

6.2 Single-tap frequency domain equalizer

The single-tap frequency domain equalizer assumes a slowly time-varying multipath channel, where each subcarrier remains orthogonal after transmission over the channel. This allows a low complexity equalization in the frequency domain, leading to a decent detection method for OTFS in slowly time-varying channels.

6.2.1 Single-tap equalizer for RCP-OTFS

We consider RCP-OTFS, where a single CP is prepended to the entire OTFS frame. Let $\mathbf{s},\ \mathbf{r} \in \mathbb{C}^{NM \times 1}$ be the time domain transmitted and received OTFS samples vectors converted from the delay-Doppler vectors $\mathbf{x}$ and $\mathbf{y}$ using (4.35). Assuming rectangular pulse shaping waveforms in the rest of the chapter, we recall the vectorized time domain input–output relation in (4.37) as

$$\boxed{\mathbf{r} = \mathbf{G} \cdot \mathbf{s} + \mathbf{w},} \tag{6.4}$$

where $\mathbf{G} \in \mathbb{C}^{NM \times NM}$ is the time domain channel matrix and its entries are given in (4.38), and $\mathbf{w} \in \mathbb{C}^{NM \times 1}$ denotes the time domain AWGN vector with σ_w^2 variance.

Assuming the case of a static (or very low mobility) multipath channel, $\mathbf{G}$ becomes a circulant matrix, similar to OFDM with NM subcarriers. Let $\check{\mathbf{s}},\ \check{\mathbf{r}} \in \mathbb{C}^{NM \times 1}$ be the time-frequency domain samples vectors obtained by the NM-point FFT (i.e., $\mathbf{F}_{MN}$) of the time domain vectors $\mathbf{s}$ and $\mathbf{r}$ using

$$\check{\mathbf{s}} = \mathbf{F}_{MN} \cdot \mathbf{s}, \quad \check{\mathbf{r}} = \mathbf{F}_{MN} \cdot \mathbf{r}. \tag{6.5}$$

Note that $\check{\mathbf{s}}$ and $\check{\mathbf{r}}$ are *different* from the transmitted time-frequency samples $\check{\mathbf{x}}$ and $\check{\mathbf{y}}$ in Section 4.4.2, which are based on the M-point FFT (i.e., $\mathbf{F}_M$) of the N blocks of $\mathbf{s}$ and $\mathbf{r}$, i.e., $\check{\mathbf{x}} = (\mathbf{I}_N \otimes \mathbf{F}_M) \cdot \mathbf{s}$ and $\check{\mathbf{y}} = (\mathbf{I}_N \otimes \mathbf{F}_M) \cdot \mathbf{r}$.

The input–output relation of (6.4) in the time-frequency domain can be written as

$$\boxed{\check{\mathbf{r}} = \check{\mathbf{G}} \cdot \check{\mathbf{s}} + \check{\mathbf{z}},} \tag{6.6}$$

where $\check{\mathbf{G}} = \mathbf{F}_{MN} \cdot \mathbf{G} \cdot \mathbf{F}_{MN}^{\dagger}$ and $\check{\mathbf{z}} = \mathbf{F}_{MN} \cdot \mathbf{w}$.

Since $\mathbf{G}$ is circulant, $\check{\mathbf{G}}$ is a diagonal matrix. The zero-forcing (ZF) estimate of the transmitted frequency domain samples is

$$\boxed{\check{\mathbf{s}}[q] = \frac{\check{\mathbf{r}}[q]}{\check{\mathbf{G}}[q,q]},} \tag{6.7}$$

where $q = 0, \ldots, NM - 1$. To avoid noise enhancement due to inverting small channel coefficients in $\check{\mathbf{G}}[q,q]$, the Fourier domain minimum mean-square error (MMSE) estimate of the frequency domain samples can be obtained as

$$\boxed{\check{\mathbf{s}}[q] = \frac{\check{\mathbf{G}}^*[q,q]\check{\mathbf{r}}[q]}{|\check{\mathbf{G}}[q,q]|^2 + \sigma_w^2},} \tag{6.8}$$

where $q = 0, \ldots, NM - 1$. The MMSE estimate can offer better performance than ZF in the case of frequency selective channels with spectral nulls. The equalized frequency domain samples vector is converted to time domain using the MN-point IFFT operation as

$$\widehat{\mathbf{s}} = \mathbf{F}_{MN}^{\dagger} \cdot \check{\mathbf{s}}, \tag{6.9}$$

which is then transformed back to the delay-Doppler domain to obtain the estimated symbols vector

$$\widehat{\mathbf{x}} = (\mathbf{I}_M \otimes \mathbf{F}_N) \cdot \mathbf{P}^{\mathrm{T}} \cdot \widehat{\mathbf{s}}. \tag{6.10}$$

6.2.2 Block-wise single-tap equalizer for CP-OTFS

In the previous section, we discussed the time-frequency domain single-tap equalizer for RCP-OTFS in static (frequency selective) multipath channels. Diagonalization of the matrix $\mathbf{G}$ of size $NM \times NM$ has a complexity of $O(2NM \log_2(NM))$. This complexity can be further reduced if a CP is added to each of the N blocks in the time domain (CP-OTFS) rather than one CP for the entire OTFS frame (RCP-OTFS). This will enable parallel operations on N circulant channel matrices of smaller size $M \times M$.

We assume static (or very low mobility) multipath channels. Recall from Chapter 4 that, for CP-OTFS, the time domain channel matrix $\mathbf{G}$ is a block diagonal matrix with only circulant blocks $\mathbf{G}_{n,0} \in \mathbb{C}^{M \times M}$ due to static (or very low mobility) channels, where subscript 0 refers to the blocks of the diagonal of $\mathbf{G}$[1] and $n = 0, \ldots, N-1$.

Let $\mathbf{s} = [\mathbf{s}_0^{\mathrm{T}}, \ldots, \mathbf{s}_{N-1}^{\mathrm{T}}]^{\mathrm{T}}$, $\mathbf{r} = [\mathbf{r}_0^{\mathrm{T}}, \ldots, \mathbf{r}_{N-1}^{\mathrm{T}}]^{\mathrm{T}}$, and $\mathbf{w} = [\mathbf{w}_0^{\mathrm{T}}, \ldots, \mathbf{w}_{N-1}^{\mathrm{T}}]^{\mathrm{T}}$, where $\mathbf{s}_n,\ \mathbf{r}_n \in \mathbb{C}^{M \times 1}$ are the transmitted and received samples vectors and $\mathbf{w}_n \in \mathbb{C}^{M \times 1}$ is the AWGN vector with σ_w^2 variance. The time domain input–output relation in (6.4) is split block-wise, for $n = 0, \ldots, N-1$, as

$$\boxed{\mathbf{r}_n = \mathbf{G}_{n,0} \cdot \mathbf{s}_n + \mathbf{w}_n,} \tag{6.11}$$

and the corresponding time-frequency domain block-wise input–output relation is

$$\boxed{\check{\mathbf{y}}_n = \check{\mathbf{H}}_{n,0} \cdot \check{\mathbf{x}}_n + \check{\mathbf{w}}_n} \tag{6.12}$$

with

$$\check{\mathbf{H}}_{n,0} = \mathbf{F}_M \cdot \mathbf{G}_{n,0} \cdot \mathbf{F}_M^{\dagger}, \tag{6.13}$$

$$\check{\mathbf{y}}_n = \mathbf{F}_M \cdot \mathbf{r}_n. \tag{6.14}$$

[1]Similarly, this applies to $\check{\mathbf{H}}_{n,0}$.

The single tap MMSE equalizer in (6.8) is modified to

$$\boxed{\check{\mathbf{x}}_n[m] = \frac{\check{\mathbf{H}}^*_{n,0}[m,m]\check{\mathbf{y}}_n[m]}{|\check{\mathbf{H}}_{n,0}[m,m]|^2 + \sigma_w^2},} \tag{6.15}$$

where

$$\check{\mathbf{y}}_n[m] = \check{\mathbf{H}}_{n,0}[m,m]\check{\mathbf{x}}_n[m] + \underbrace{\sum_{m'=0,m'\neq m'}^{M-1} \check{\mathbf{H}}_{n,0}[m,m']\check{\mathbf{x}}_n[m'] + \check{\mathbf{w}}_n[m]}_{\text{ICI}}$$

for $m = 0, \ldots, M-1, n = 0, \ldots, N-1$, contains the intercarrier interference (ICI) term, and $\check{\mathbf{w}}_n[m]$ is the complex AWGN with zero mean and σ_w^2 variance. The estimated time-frequency domain samples vector is transformed to the time domain using the M-point IFFT operation as

$$\widehat{\mathbf{s}}_n = \mathbf{F}_M^\dagger \cdot \check{\mathbf{x}}_n. \tag{6.16}$$

The estimated delay-Doppler domain symbols vector is obtained as

$$\widehat{\mathbf{x}} = (\mathbf{I}_M \otimes \mathbf{F}_N) \cdot \mathbf{P}^{\mathrm{T}} \cdot \widehat{\mathbf{s}}, \tag{6.17}$$

where $\widehat{\mathbf{s}} = [\widehat{\mathbf{s}}_0^{\mathrm{T}}, \ldots, \widehat{\mathbf{s}}_{N-1}^{\mathrm{T}}]^{\mathrm{T}}$. Note that the steps (6.16) and (6.17) constitute the SFFT operation to convert the equalized samples from the frequency-time domain to the delay-Doppler domain.

As shown in (6.15), if there is no noise, the performance of the single-tap equalizer depends on the power ratio between the ICI term and the diagonal element $\check{\mathbf{H}}_{n,0}[m,m]$. When the Doppler spread increases as in high mobility wireless channels, the power of the ICI term (from the off-diagonal elements $\check{\mathbf{H}}_{n,0}[m,m']$) increases, thereby leading to performance degradation.

6.2.3 Complexity

As discussed above, for RCP-OTFS, the matrix diagonalization requires $2NM\log_2(NM)$ complex multiplications, since $\mathbf{G}$ is a circulant matrix. For CP-OTFS, the diagonalization requires only $2NM\log_2(M)$ complex multiplications, since $\mathbf{G}$ is a block diagonal matrix with circulant blocks $\mathbf{G}_{n,0}$. Further, the common steps in (6.10) and (6.17) for RCP-OTFS and CP-OTFS, respectively, need $MN\log_2(N)$ complex multiplications. Hence, the overall complexity for the single-tap equalizer for RCP-OTFS and CP-OTFS are $NM(3\log_2(N) + 2\log_2(M))$ and $NM(\log_2(N) + 2\log_2(M))$. Here, CP-OTFS has slightly lower complexity due to its block-wise M-point FFT operation rather than a MN-point FFT in RCP-OTFS.

6.3 Linear minimum mean-square error detection

In the previous section, we discussed the single-tap equalizer, which is similar to the commonly used detector for OFDM. Even though the single-tap equalizer has very low complexity, it works well only for static or very low mobility wireless channels. This is because the Doppler spread caused by high mobility introduces ICI interference, leading to performance degradation. Hence, in this section, we will discuss the well-known linear minimum mean-square error (LMMSE) detector for OTFS, which can offer good performance in both static and high mobility channels.

6.3.1 Delay-Doppler domain LMMSE detection

Recalling the input–output relation in (6.2), the LMMSE estimate of $\mathbf{x}$ is given as

$$\boxed{\widehat{\mathbf{x}} = (\mathbf{H}^{\dagger} \cdot \mathbf{H} + \sigma_w^2 \mathbf{I}_{MN})^{-1} \cdot \mathbf{H}^{\dagger} \cdot \mathbf{y}.} \tag{6.18}$$

The operation in (6.18) requires inverting an $NM \times NM$ matrix, which is of high complexity, even if the matrix $\mathbf{H}$ is sparse. As discussed in Section 6.1, the delay-Doppler channel matrix $\mathbf{H}$ has $S \geq P$ nonzero elements in each row, corresponding to the number of discrete delay-Doppler paths in the delay-Doppler grid. However, the time domain channel matrix $\mathbf{G}$ has just $|\mathcal{L}|$ nonzero elements in each row. If there is one integer Doppler per delay bin, then $|\mathcal{L}| = P = S$; otherwise, $|\mathcal{L}| < P < S$. The increased sparsity of the time domain matrix $\mathbf{G}$ can be used to further reduce complexity of the LMMSE detection in the time domain, as discussed in the next subsection.

6.3.2 Time domain LMMSE detection

Recalling the time domain input–output relation of RCP/RZP-OTFS in (6.4), the LMMSE estimate of $\mathbf{s}$ is given as

$$\boxed{\widehat{\mathbf{s}} = (\mathbf{G}^{\dagger} \cdot \mathbf{G} + \sigma_w^2 \mathbf{I}_{MN})^{-1} \cdot \mathbf{G}^{\dagger} \cdot \mathbf{r}.} \tag{6.19}$$

The estimated time domain samples are then transformed back to the delay-Doppler domain to obtain the estimated information symbols using the relation in (6.17). The operation in (6.19) requires inverting an $NM \times NM$ matrix. However, if we adopt CP/ZP-OTFS, then the LMMSE computation can be performed block-wise by inverting N complex submatrices of size $M \times M$ (see (6.20)). Recalling the block-wise input–output relation in (6.11) yields the following LMMSE estimate:

$$\boxed{\widehat{\mathbf{s}}_n = (\mathbf{G}_{n,0}^{\dagger} \cdot \mathbf{G}_{n,0} + \sigma_w^2 \mathbf{I}_M)^{-1} \mathbf{G}_{n,0}^{\dagger} \cdot \mathbf{r}_n} \tag{6.20}$$

for $n = 0, \ldots, N-1$. Finally, the block estimates are converted back to obtain the delay-Doppler information symbols similar to (6.17).

The readers can find the MATLAB® implementation of the delay-Doppler and time domain LMMSE detection in MATLAB codes 13 and 14 of Appendix C.

6.3.3 Complexity

The LMMSE methods of (6.18) and (6.19) in the delay-Doppler domain and time domain, respectively, require inverting an $NM \times NM$ matrix, thereby incurring a complexity of $O((NM)^3)$. Note that the cubic complexity can be reduced by taking advantage of the banded and sparse structure of $\mathbf{H}$ and $\mathbf{G}$, as well as the fact that the time domain matrix $\mathbf{G}$ is sparser than $\mathbf{H}$, since $|\mathcal{L}| \leq P \leq S$.

For CP/ZP-OTFS, the LMMSE detection can be performed block-wise for N blocks (see (6.20)), which requires inverting N submatrices of size $M \times M$, followed by transforming the time domain estimates to the delay-Doppler domain. Hence, such detection has an overall complexity of $O(NM^3 + NM \log_2 N))$, which is significantly lower than for the LMMSE methods in (6.18) and (6.19). In addition, the sparsity and banded nature of these submatrices can be exploited to further reduce detection complexity.

6.4 Message passing detection

In this section, we present the state-of-the-art message passing (MP) detection for OTFS. The LMMSE detection can offer good performance in doubly selective channels but at the cost of high complexity, as discussed above. Single tap equalization however requires the lowest complexity but its performance degrades with increasing Doppler shifts. The MP detection can offer better performance than the LMMSE detection with much lower complexity and can be applied to all variants of OTFS.

6.4.1 Message passing detection algorithm

Consider the vectorized delay-Doppler input–output relation in (6.2), where $\mathbf{y}, \ \mathbf{z} \in \mathbb{C}^{NM \times 1}$ are the received and noise vectors with elements $y[d]$ and $z[d]$, $\mathbf{H} \in \mathbb{C}^{NM \times NM}$ is the delay-Doppler channel matrix with elements $H[d, c]$, and $\mathbf{x} \in \mathbb{C}^{NM \times 1}$ is the information vector with elements $x[c] \in \mathbb{A}$, where $d, c \in [0, \ NM-1]$ and $\mathbb{A} = \{a_1, \ldots, a_Q\}$ represents a modulation alphabet of size Q. Let $\mathcal{I}(d)$ and $\mathcal{J}(c)$ denote the sets of indexes with nonzero elements in the d-th row and c-th column, respectively, and $|\mathcal{I}(d)| = |\mathcal{J}(c)| = S$ for all rows and columns.

Based on (6.2), we model the system as a sparsely connected factor graph with NM variable nodes corresponding to $\mathbf{x}$ and NM observation

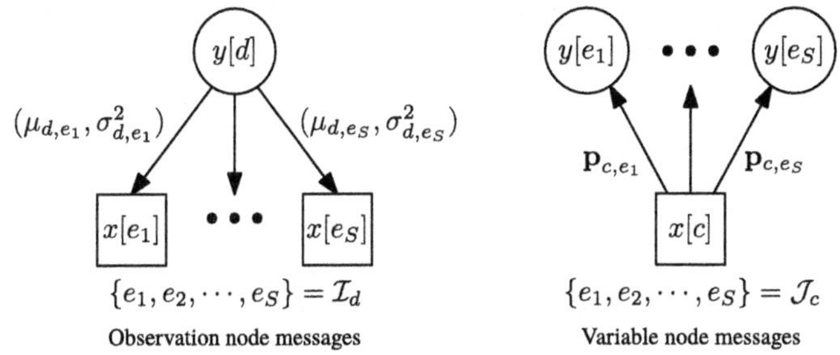

FIGURE 6.1 Messages exchanged in the factor graph.

nodes corresponding to $\mathbf{y}$. In this factor graph, each observation node $y[d]$ is connected to the set of S variable nodes $\{x[c], c \in \mathcal{I}(d)\}$. Similarly, each variable node $x[c]$ is connected to the set of S observation nodes $\{y[d], d \in \mathcal{J}(c)\}$.

From (6.2), the joint maximum a posteriori probability (MAP) detection rule for estimating the transmitted signals is given by

$$\widehat{\mathbf{x}} = \arg \max_{\mathbf{x} \in \mathbb{A}^{NM \times 1}} \Pr\left(\mathbf{x} \mid \mathbf{y}, \mathbf{H}\right),$$

which has an exponential complexity in NM. Since the joint MAP detection can be intractable for practical values of N and M, we consider the symbol-by-symbol MAP detection rule for $c = 0, \ldots, NM - 1$

$$\begin{aligned}\widehat{x}[c] &= \arg\max_{a_j \in \mathbb{A}} \Pr\left(x[c] = a_j \mid \mathbf{y}, \mathbf{H}\right) \\ &= \arg\max_{a_j \in \mathbb{A}} \frac{1}{Q} \Pr\left(\mathbf{y} \mid x[c] = a_j, \mathbf{H}\right) \quad (6.21) \\ &\approx \arg\max_{a_j \in \mathbb{A}} \prod_{d \in \mathcal{J}_c} \Pr\left(y[d] \mid x[c] = a_j, \mathbf{H}\right). \quad (6.22)\end{aligned}$$

In (6.21), we assume all the transmitted symbols $a_j \in \mathbb{A}$ are equally likely and in (6.22), we assume that $y[d]$ for all $d \in \mathcal{J}_c$ are approximately independent for a given $x[c]$, due to the sparsity of $\mathbf{H}$. Then we can assume that the interference terms $\zeta^{(i)}_{d,c}$ defined in (6.23) are independent for a given c.

To solve the approximate symbol-by-symbol MAP detection in (6.22), we present an MP detector that has a linear complexity in NM. For each $y[d]$, the variable $x[c]$ is isolated from the other interference terms, which are approximated as Gaussian noise with the easily computable mean and variance.

In the MP algorithm, the mean and variance of the interference terms are used as messages from observation nodes to variable nodes. In addition, the message passed from a variable node $x[c]$ to the observation

Algorithm 1: MP algorithm for OTFS symbol detection.

Input: Channel matrix $\mathbf{H}$, received signal $\mathbf{y}$
Initialize: pmf $\mathbf{p}_{c,d}^{(0)} = 1/Q, c = 0, \cdots, NM-1, d \in \mathcal{J}(c)$.
for *i=1:max iterations* **do**
  Observation nodes $y[d]$ compute the means $\mu_{d,c}^{(i)}$ and variances $(\sigma_{d,c}^{(i)})^2$ of Gaussian random variables $\zeta_{d,c}^{(i)}$ using $\mathbf{p}_{c,d}^{(i-1)}$ and pass them to variables nodes $x[c], c \in \mathcal{I}(d)$.
  Variable nodes $x[c]$ update $\mathbf{p}_{c,d}^{(i)}$ using $\mu_{d,c}^{(i)}$, $(\sigma_{d,c}^{(i)})^2$, and $\mathbf{p}_{c,d}^{(i-1)}$ and pass them to observation nodes $y[d], d \in \mathcal{J}(c)$.
  Compute convergence indicator $\eta^{(i)}$.
  Update the decision on the transmitted symbols $\widehat{x}[c], c = 0, \ldots, NM-1$ if needed.
  if (Stopping criteria satisfied) **then EXIT**
end
Output: The decision on transmitted symbols $\widehat{x}[c]$.

nodes $y[d], d \in \mathcal{J}(c)$, is the probability mass function (pmf) of the alphabet $\mathbf{p}_{c,d} = \{p_{c,d}(a_j)|a_j \in \mathbb{A}\}$. Fig. 6.1 shows the connections and the messages passed between the observation and variable nodes. The MP algorithm is given in Algorithm 1, where the steps in the i-th iteration are detailed below.

1. **Observation nodes** $y[d] \rightarrow$ **variable nodes** $x[c], c \in \mathcal{I}(d)$:
Mean $\mu_{d,c}^{(i)}$ and variance $(\sigma_{d,c}^{(i)})^2$ of the interference, approximately modeled as a Gaussian random variable $\zeta_{d,c}^{(i)}$ defined as

$$y[d] = x[c]H[d,c] + \underbrace{\sum_{e \in \mathcal{I}(d), e \neq c} x[e]H[d,e] + z[d]}_{\zeta_{d,c}^{(i)}}, \tag{6.23}$$

can be computed as

$$\mu_{d,c}^{(i)} = \sum_{e \in \mathcal{I}(d), e \neq c} \sum_{j=1}^{Q} p_{e,d}^{(i-1)}(a_j) a_j H[d,e], \tag{6.24}$$

and

$$(\sigma_{d,c}^{(i)})^2 = \sum_{e \in \mathcal{I}(d), e \neq c} \left(\sum_{j=1}^{Q} p_{e,d}^{(i-1)}(a_j) |a_j|^2 |H[d,e]|^2 v \right.$$

$$- \left| \sum_{j=1}^{Q} p_{e,d}^{(i-1)}(a_j) a_j H[d,e] \right|^2 \Bigg) + \sigma^2. \tag{6.25}$$

2. **Variable nodes** $x[c] \rightarrow$ **observation nodes** $y[d], d \in \mathcal{J}(c)$:
The pmf vector $\mathbf{p}_{c,d}^{(i)}$ can be updated as

$$p_{c,d}^{(i)}(a_j) = \Delta \cdot \tilde{p}_{c,d}^{(i)}(a_j) + (1-\Delta) \cdot p_{c,d}^{(i-1)}(a_j),\ a_j \in \mathbb{A}, \tag{6.26}$$

where $\Delta \in (0, 1]$ is the *damping factor* used to improve the performance by controlling the convergence speed and

$$\begin{aligned} \tilde{p}_{c,d}^{(i)}(a_j) &\propto \prod_{e \in \mathcal{J}(c), e \neq d} \Pr\left(y[e] \middle| x[c] = a_j, \mathbf{H}\right) \\ &= \prod_{e \in \mathcal{J}(c), e \neq d} \frac{\xi^{(i)}(e,c,j)}{\sum_{k=1}^{Q} \xi^{(i)}(e,c,k)}, \end{aligned} \tag{6.27}$$

where

$$\xi^{(i)}(e,c,k) = \exp\left(\frac{-\left| y[e] - \mu_{e,c}^{(i)} - H_{e,c} a_k \right|^2}{(\sigma_{e,c}^{(i)})^2} \right). \tag{6.28}$$

3. **Convergence indicator** $\eta^{(i)}$: Compute $\eta^{(i)}$ as

$$\eta^{(i)} = \frac{1}{NM} \sum_{c=1}^{NM} \mathbb{I}\left(\max_{a_j \in \mathbb{A}} p_c^{(i)}(a_j) \geq 1 - \gamma \right), \tag{6.29}$$

for some small $\gamma > 0$, where

$$p_c^{(i)}(a_j) = \prod_{e \in \mathcal{J}(c)} \frac{\xi^{(i)}(e,c,j)}{\sum_{k=1}^{Q} \xi^{(i)}(e,c,k)} \tag{6.30}$$

and $\mathbb{I}(\cdot)$ is an indicator function which gives a value of 1 if the expression in the argument is true and 0 otherwise.
4. **Update decision**: If $\eta^{(i)} > \eta^{(i-1)}$, then we update the decision of the transmitted symbol as

$$\widehat{x}[c] = \arg\max_{a_j \in \mathbb{A}} p_c^{(i)}(a_j), \quad c = 0, \cdots, NM-1. \tag{6.31}$$

5. **Stopping criteria.** The MP algorithm stops when at least one of the following conditions is satisfied:
   a. $\eta^{(i)} = 1$,

b. $\eta^{(i)} < \eta^{(i^*)} - \epsilon$, where $i^* \in \{1, \cdots, (i-1)\}$ is the iteration index for which $\eta^{(i^*)}$ is maximum,
c. the maximum number n_{iter} of iterations is reached.

We set $\epsilon = 0.2$ to disregard small fluctuations of η. Here, the first condition occurs in the best case, where all the symbols have converged. The second condition is useful to stop the algorithm if the current iteration provides a worse decision than the best one in previous iterations.

6.4.2 Complexity

The complexity of each iteration involves the computation of (6.24), (6.25), (6.26), (6.29), and (6.31), each of which is of a complexity order $O(NMSQ)$. This is because, when computing (6.27), we need to find $p_c^{(i)}(a_j)$ in (6.30), which is of complexity $O(NMQ)$, and then obtain (6.27) by dividing (6.30) with the term related to $e = d$ for all d, requiring complexity $O(S)$ for each c.

Furthermore, the computation of (6.29) and (6.31) requires to find the maximum element out of Q elements for each c. Since (6.30) is already computed, finding the maximum element requires $O(Q)$ complexity for each c, leading to an overall complexity of $O(NMQ)$.

Hence, the overall complexity order is $O(\mathrm{n}_{\text{iter}}NMSQ)$. In simulations, we observed that the algorithm converges typically with $\mathrm{n}_{\text{iter}} \leq 20$, when using a damping factor $\Delta = 0.7$. We conclude that the analysis on inter-Doppler interference including the smart approximation of such interference to exploit the sparsity of the delay-Doppler channel is a key factor in reducing the detection complexity. The memory requirement is dominated by the storage of $2NMSQ$ real values for $\mathbf{p}_{c,d}^{(i)}$ and $\mathbf{p}_{c,d}^{(i-1)}$. In addition, we have the messages $(\mu_{d,c}^{(i)}, (\sigma_{d,c}^{(i)})^2)$, requiring NMS complex values and NMS real values, respectively.

6.5 Maximum-ratio combining detection

In this section, we discuss the maximum-ratio combining (MRC) detection, which offers similar performance to that of the MP detection but at much lower complexity. The proposed MRC detection method is motivated by the traditional rake receiver in code division multiple access (CDMA) systems, since OTFS can be interpreted as a two-dimensional CDMA with information symbols spreading in both time and frequency.

In direct sequence CDMA operating in a multipath fading channel, a rake receiver combines the delayed components (or echoes) of the transmitted symbols, which are extracted using matched filters tuned to the respective delay shifts. Similarly, in OTFS, the channel impaired signal components received at $|\mathcal{L}|$ different delay branches can be extracted and

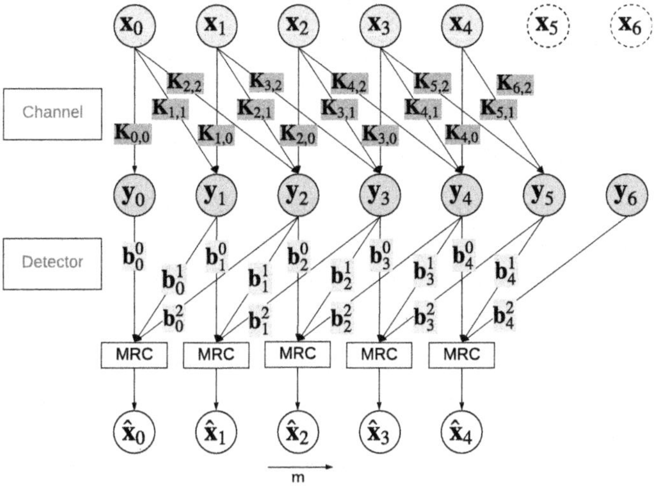

FIGURE 6.2 MRC delay-Doppler domain operation for $M' = 7$, $M = 5$, and the set of discrete delay indices $\mathcal{L} = \{0, 1, 2\}$. [Copyright permission obtained.]

coherently combined using diversity combining techniques such as MRC (see Fig. 6.2) to improve the signal-to-noise ratio (SNR) of the accumulated signal.

It is well known that MRC is an optimal combiner for independent AWGN channels. Nevertheless, it is also optimal with correlated channels if the branch weights are selected according to the signal amplitude-to-noise power ratios of the correlated branches. For OTFS, the noise plus interference (NPI) in each branch is correlated, since it depends on the channel, which introduces interference across the branches. In the MRC detection, we can cancel the estimated interference in the branches selected for combining and iteratively improve the post-MRC signal-to-interference-plus-noise ratio (SINR).

In the following, we present the *delay-Doppler* MRC detection and the reduced complexity *delay-time* MRC detection for ZP-OTFS. These MRC detection methods can be easily extended to the other OTFS variants.

For ease of derivation, we consider N ZPs, each of length $L_{\text{ZP}} = l_{\max}$, to be part of an extended OTFS frame of size $M'N$, where $M' = M + l_{\max}$. At the receiver, the ZPs are not discarded, but utilized for the MRC detection.

6.5.1 Delay-Doppler domain MRC detection

With a ZP in each block, we set the transmitted and received symbol vectors as

$$\mathbf{x} = [\mathbf{x}_0^T, \ldots, \mathbf{x}_{M'-1}^T]^T,$$

$$\mathbf{y} = [\mathbf{y}_0^T, \ldots, \mathbf{y}_{M'-1}^T]^T, \tag{6.32}$$

with subvectors $\mathbf{x}_m, \mathbf{y}_m \in \mathbb{C}^{N\times 1}$ for $m = 0, \ldots, M' - 1$, and the zero padding

$$\mathbf{x}_m = \mathbf{0} \quad \text{for} \quad m = M, \ldots, M' - 1.$$

Note that the symbol vectors in (6.32) include the ZPs, which differ from those in (6.1), which excluded the ZPs. With the effect of the ZP, the delay-Doppler input–output relation in (6.3) can be modified as

$$\mathbf{y}_{m'} = \sum_{l'\in\mathcal{L}} \mathbf{K}_{m',l'} \cdot \mathbf{x}_{m'-l'} + \mathbf{z}_{m'} \tag{6.33}$$

for $m' = 0, \ldots, M' - 1, 0 \le l' \le l_{\max}, 0 \le m' - l' \le M - 1$, where $\mathbf{K}_{m',l'}$ are the submatrices of the extended delay-Doppler channel matrix $\mathbf{H} \in \mathbb{C}^{NM'\times NM'}$ and $\mathbf{z}_{m'}$ are the AWGN vectors.

Next, we conduct the detection procedure iteratively for estimating $\mathbf{x}_m$ only for $m = 0, \ldots, M - 1$, since $\mathbf{x}_m = \mathbf{0}$ for $m = M, \ldots, M' - 1$ are the ZP. For ease of notation, we *omit* the iteration index in the following steps.

Let $\mathbf{b}_m^l \in \mathbb{C}^{N\times 1}$, where $m = 0, \ldots, M - 1$ and $l \in \mathcal{L}$, be the channel impaired signal component of $\mathbf{x}_m$ in the received vectors $\mathbf{y}_{m+l}$ after removing the interference of other transmitted symbol vectors $\mathbf{x}_k$ for $k \neq m$. Assuming we have the estimates of $\mathbf{x}_m$ from previous iterations, we can write $\mathbf{b}_m^l$ for $l \in \mathcal{L}$ as

$$\mathbf{b}_m^l = \mathbf{y}_{m+l} - \underbrace{\sum_{l'\in\mathcal{L}, l'\neq l} \mathbf{K}_{m+l,l'} \cdot \widehat{\mathbf{x}}_{m+l-l'}}_{\text{interdelay interference}}, \tag{6.34}$$

where $\widehat{\mathbf{x}}_{m+l-l'} = \mathbf{0}$ for $m + l - l' < 0$ and $\mathbf{K}_{m+l,l'}$ are the submatrices of the extended delay-Doppler channel matrix $\mathbf{H} \in \mathbb{C}^{NM'\times NM'}$.

Substituting (6.33) in (6.34) with $m' = m + l$ yields

$$\mathbf{b}_m^l = \mathbf{K}_{m+l,l} \cdot \mathbf{x}_m + \underbrace{\sum_{l'\in\mathcal{L}, l'\neq l} \mathbf{K}_{m+l,l'} \cdot (\mathbf{x}_{m+l-l'} - \widehat{\mathbf{x}}_{m+l-l'}) + \mathbf{z}_{m+l}}_{\text{interference and noise}}. \tag{6.35}$$

In this scheme, instead of estimating $\widehat{\mathbf{x}}_m$ separately from each equation in (6.35), we perform maximal ratio combining of the channel impaired components $\mathbf{b}_m^l$ as

$$\mathbf{c}_m = \left(\sum_{l\in\mathcal{L}} \mathbf{K}_{m+l,l}^\dagger \cdot \mathbf{K}_{m+l,l}\right)^{-1} \cdot \left(\sum_{l\in\mathcal{L}} \mathbf{K}_{m+l,l}^\dagger \cdot \mathbf{b}_m^l\right) = \mathbf{D}_m^{-1} \cdot \mathbf{g}_m, \tag{6.36}$$

where $\mathbf{c}_m \in \mathbb{C}^{N\times 1}$ is the vector output of the maximal ratio combiner and

$$\mathbf{D}_m = \sum_{l\in\mathcal{L}} \mathbf{K}_{m+l,l}^{\dagger} \cdot \mathbf{K}_{m+l,l}, \tag{6.37}$$

$$\mathbf{g}_m = \sum_{l\in\mathcal{L}} \mathbf{K}_{m+l,l}^{\dagger} \cdot \mathbf{b}_m^{l}, \tag{6.38}$$

followed by the symbol-by-symbol maximum likelihood detection (MLD), resulting in the hard estimates given by

$$\widehat{\mathbf{x}}_m[n] = \arg\min_{a_j\in\mathcal{Q}} \left|a_j - \mathbf{c}_m[n]\right|, \tag{6.39}$$

where a_j is a signal from the QAM alphabet $\mathbb{A}$ of size Q, with $j = 1, \ldots, Q$ and $n = 0, \ldots, N-1$. Let $\mathcal{D}(\cdot)$ denote the decision on the estimate $\mathbf{c}_m$ in every iteration such that $\widehat{\mathbf{x}}_m = \mathcal{D}(\mathbf{c}_m)$. Hard decision function $\mathcal{D}(\cdot)$ is given by the MLD criterion in (6.39).

Once we update the estimate $\widehat{\mathbf{x}}_m$, we increase m and repeat the same procedure to estimate all M symbol vectors $\widehat{\mathbf{x}}_m$ using the updated estimates[2] of the previously decoded ones in the form of a decision feedback equalizer (DFE) (see Fig. 6.2). Note that the DFE action leads to sequential updates, whereas alternatively using only the previous iteration estimates leads to parallel updates. We later verify experimentally that parallel updates result in slower convergence.

Algorithm 2 provides the detailed delay-Doppler MRC operation.

6.5.2 Complexity

Delay-Doppler implementation complexity

The core iterative operations of the delay-Doppler MRC detection can be identified as steps 7, 9, and 10 of Algorithm 2. Due to the operations involving matrix–vector products, the *three* steps together incur the complexity of order $O(N^2|\mathcal{L}|^2)$ per iteration per symbol vector. The calculation of $\mathbf{D}_m$ needs to be performed only once and can be reused in each iteration and requires $O(N^3)$ complexity per symbol vector. Overall the detection complexity is given as $O(M(N^3 + \mathrm{n_{iter}} N^2|\mathcal{L}|))$, where $\mathrm{n_{iter}}$ is the number of iterations.

This complexity is significantly lower than that of LMMSE and comparable to that of the MP algorithm, but still much higher than the desired complexity. However the exponential dependency on N can be mitigated by taking advantage of some special properties of the OTFS channel matrix.

[2] Alternatively, a soft estimate can also be used in conjunction with an outer coding scheme as described in Section 6.6.

Algorithm 2: MRC in delay-Doppler domain.

Input: $\mathbf{x}_m = \mathbf{0}_N \quad \text{for } m = 0, \ldots, M-1$
Input: $\mathbf{H} \in \mathbb{C}^{NM' \times NM'}$ and $\mathbf{y}_m$ for $m = 0, \ldots, M'-1$
for *i=1:max iterations* **do**
  for $m = 0 : M-1$ **do**
    $\mathbf{D}_m = \sum_{l \in \mathcal{L}} \mathbf{K}^{\dagger}_{m+l,l} \mathbf{K}_{m+l,l}$
    for $l \in \mathcal{L}$ **do**
      $\mathbf{b}^l_m = \mathbf{y}_{m+l} - \sum_{l' \neq l} \mathbf{K}_{m+l,l'} \cdot \widehat{\mathbf{x}}_{m+l-l'}$
    end
    $\mathbf{g}_m = \sum_{l \in \mathcal{L}} \mathbf{K}^{\dagger}_{m+l,l} \cdot \mathbf{b}^l_m$
    $\mathbf{c}_m = \mathbf{D}^{-1}_m \cdot \mathbf{g}_m$
    $\widehat{\mathbf{x}}_m = \mathcal{D}(\mathbf{c}_m) \quad (\text{or} \quad \widehat{\mathbf{x}}_m = \mathbf{c}_m)$
  end
end
Output: $\widehat{\mathbf{x}}_m$

Initial step complexity

In Algorithm 2, the initial computations include generating all entries of $\mathbf{H}$, which require computing $\boldsymbol{\nu}_{m,l}$ for $m = 0, \ldots, M'-1$ and $l \in \mathcal{L}$. Assuming the integer delay-Doppler channel parameters (h_i, l_i, k_i) are known for $i = 1, 2, \ldots, P$, the channel Doppler spread vectors $\boldsymbol{\nu}_{m,l}$ can be easily computed using the relations given in (2.13) and (4.81). Specifically, for the delay-Doppler domain MRC operation in Algorithm 2, let K_l be the number of nonzero channel coefficients in each vector $\boldsymbol{\nu}_{m,l}$ (or paths with different Doppler shift in the same delay bin $l \in \mathcal{L}$) such that the total number of channel coefficients or propagation paths as seen by the OTFS receiver is $P = \sum_{l \in \mathcal{L}} K_l$. The number of complex multiplications required to compute $\boldsymbol{\nu}_{m,l}$ for $m = 0, \ldots, M'-1$ and $l \in \mathcal{L}$, using (4.81), is $M \sum_{l \in \mathcal{L}} K_l = M'P$. Hence, the OTFS channel matrix $\mathbf{H}$ can be generated in $M'P$ complex multiplications.

6.5.3 Reduced complexity delay-time domain implementation

In (6.34), for each symbol vector $\mathbf{x}_m$, we need to compute all vectors $\mathbf{b}^l_m$ for $l \in \mathcal{L}$. This operation requires $|\mathcal{L}|(|\mathcal{L}|-1)$ products between $\mathbf{K}_{m,l}$ and $\widehat{\mathbf{x}}_{m-l}$. These operations can be reduced by using the following approach.

Let us define the *residual error plus noise* (REPN) term using symbol vectors estimates $\widehat{\mathbf{x}}_{m-l}$ for $l \in \mathcal{L}$ as

$$\Delta \mathbf{y}_m = \mathbf{y}_m - \sum_{l \in \mathcal{L}} \mathbf{K}_{m,l} \cdot \widehat{\mathbf{x}}_{m-l}. \tag{6.40}$$

To illustrate the algorithm, we now include the *iteration index* in the following equations. We can set the initial estimate $\widehat{\mathbf{x}}_{m-l}^{(0)} = \mathbf{0}_N$ and thus we obtain $\Delta\mathbf{y}_m^{(0)} = \mathbf{y}_m$ for all m. Alternatively, the initial estimate can be computed as described in Section 6.5.5.

Note that symbol vectors estimates are obtained in increasing order $m = 0, \ldots, M-1$. Therefore, for estimating $\mathbf{x}_m$ in the i-th iteration, only the past estimated symbol vectors $[\widehat{\mathbf{x}}_0^{(i)}, \ldots, \widehat{\mathbf{x}}_{m-1}^{(i)}]$ are used.

From (6.34), we rewrite $\mathbf{b}_m^l$ for the i-th iteration as

$$(\mathbf{b}_m^l)^{(i)} = \Delta\mathbf{y}_{m+l}^{(i)} + \mathbf{K}_{m+l,l} \cdot \widehat{\mathbf{x}}_m^{(i-1)}.$$

Substituting this into (6.38) yields

$$\begin{aligned}\mathbf{g}_m^{(i)} &= \sum_{l\in\mathcal{L}} \mathbf{K}_{m+l,l}^{\dagger} \cdot \Delta\mathbf{y}_{m+l}^{(i)} + \underbrace{\left(\sum_{l\in\mathcal{L}} \mathbf{K}_{m+l,l}^{\dagger} \cdot \mathbf{K}_{m+l,l}\right)}_{\mathbf{D}_m} \cdot \widehat{\mathbf{x}}_m^{(i-1)} \\ &= \sum_{l\in\mathcal{L}} \mathbf{K}_{m+l,l}^{\dagger} \cdot \Delta\mathbf{y}_{m+l}^{(i)} + \mathbf{D}_m \cdot \widehat{\mathbf{x}}_m^{(i-1)}. \end{aligned} \tag{6.41}$$

Substituting (6.41) into (6.36) yields the MRC output at the i-th iteration as

$$\mathbf{c}_m^{(i)} = \mathbf{D}_m^{-1} \cdot \mathbf{g}_m^{(i)} = \widehat{\mathbf{x}}_m^{(i-1)} + \mathbf{D}_m^{-1} \cdot \Delta\mathbf{g}_m^{(i)}, \tag{6.42}$$

where

$$\Delta\mathbf{g}_m^{(i)} = \sum_{l\in\mathcal{L}} \mathbf{K}_{m+l,l}^{\dagger} \cdot \Delta\mathbf{y}_{m+l}^{(i)}, \tag{6.43}$$

which is the maximal ratio combination of the REPNs in all the delay branches ($\mathbf{y}_{m+l}$ for $l \in \mathcal{L}$) having a component of $\mathbf{x}_m$ in them.

In the i-th iteration, for every estimated symbol vector $\mathbf{x}_m$, there are $|\mathcal{L}|$ REPN vectors $\Delta\mathbf{y}_{m+l}^{(i)}$ that need to be updated, each one costing $|\mathcal{L}|$ matrix–vector products for a total of $|\mathcal{L}|^2$ matrix–vector products. However, the complexity of (6.40) can be reduced by storing and updating the initial REPN vectors $\Delta\mathbf{y}_m^{(0)}$. Then the vectors $\Delta\mathbf{y}_{m+l}^{(i)}$ having a component of the most recently estimated symbol vector are updated as

$$\Delta\mathbf{y}_{m+l}^{(i)} \leftarrow \Delta\mathbf{y}_{m+l}^{(i)} - \mathbf{K}_{m+l,l} \cdot (\widehat{\mathbf{x}}_m^{(i)} - \widehat{\mathbf{x}}_m^{(i-1)}). \tag{6.44}$$

Note that we can replace $\widehat{\mathbf{x}}_m^{(i)}$ by the soft estimate $\mathbf{c}_m^{(i)}$ in the above equation. The number of matrix–vector products required to compute $\Delta\mathbf{y}_m^{(i)}$ has now been reduced from $|\mathcal{L}|^2$ in (6.40) to $|\mathcal{L}|$ in (6.44).

Moreover, as described in Table 4.2, the matrix–vector products in (6.43) and (6.44) are products between circulant matrices $\mathbf{K}_{m,l} \in \mathbb{C}^{N\times N}$ and column vectors $\mathbf{x}_m$ or $\Delta\mathbf{y}_m \in \mathbb{C}^{N\times 1}$, which can be converted in the delay-time domain to element-wise products of vectors $\tilde{\boldsymbol{\nu}}_{m,l} \circ \tilde{\mathbf{x}}_m$ or $\tilde{\boldsymbol{\nu}}_{m,l} \circ \Delta\tilde{\mathbf{y}}_m$, respectively, where

$$\tilde{\boldsymbol{\nu}}_{m,l} = \mathbf{F}_N^{\dagger} \cdot \boldsymbol{\nu}_{m,l} \tag{6.45}$$

denotes the Doppler spread vector in the delay-time domain. This results in a complexity of N complex multiplications. Eqs. (6.42), (6.43), and (6.44) can now be written in corresponding delay-time domain as

$$\tilde{\mathbf{c}}_m^{(i)} = \tilde{\mathbf{x}}_m^{(i-1)} + \Delta\tilde{\mathbf{g}}_m^{(i)} \oslash \tilde{\mathbf{d}}_m, \tag{6.46}$$

$$\Delta\tilde{\mathbf{g}}_m^{(i)} = \sum_{l\in\mathcal{L}} \tilde{\boldsymbol{\nu}}_{m+l,l}^{*} \circ \Delta\tilde{\mathbf{y}}_{m+l}^{(i)}, \tag{6.47}$$

$$\Delta\tilde{\mathbf{y}}_{m+l}^{(i)} \leftarrow \Delta\tilde{\mathbf{y}}_{m+l}^{(i)} - \tilde{\boldsymbol{\nu}}_{m+l,l} \circ (\tilde{\mathbf{x}}_m^{(i)} - \tilde{\mathbf{x}}_m^{(i-1)}), \tag{6.48}$$

where we can replace $\tilde{\mathbf{x}}_m^{(i)}$ by the soft estimate $\tilde{\mathbf{c}}_m^{(i)}$ in the above equation and

$$\tilde{\mathbf{d}}_m = \sum_{l\in\mathcal{L}} \tilde{\boldsymbol{\nu}}_{m+l,l}^{*} \circ \tilde{\boldsymbol{\nu}}_{m+l,l}, \tag{6.49}$$

which can be computed with only $N|\mathcal{L}|$ complex multiplications. The detection process stops when the norm of the overall REPN error $\|\Delta\tilde{\mathbf{y}}\|$ becomes nondecreasing, where $\Delta\tilde{\mathbf{y}} = \Delta\tilde{\mathbf{y}} = [\Delta\tilde{\mathbf{y}}_0^{\mathrm{T}}, \cdots, \Delta\tilde{\mathbf{y}}_{M'-1}^{\mathrm{T}}]^{\mathrm{T}}$ becomes nonincreasing. Algorithm 3 provides the detailed delay-time domain MRC operation.

6.5.4 Complexity

Delay-time implementation complexity

In Algorithm 3, the overall complexity per iteration is measured in terms of the number of complex multiplications. Computing $\Delta\tilde{\mathbf{g}}_m^{(i)}$, $\tilde{\mathbf{c}}_m^{(i)}$, and $\Delta\tilde{\mathbf{y}}_m^{(i)}$ for all symbol vectors requires $M(2|\mathcal{L}|+1)N$ complex multiplications[3] and therefore the complexity per iteration is of the order $O(NM|\mathcal{L}|)$. The redundant FFT computations can be avoided by storing $\tilde{\boldsymbol{\nu}}_{m,l}$, the M initial symbol vector estimates $\tilde{\mathbf{x}}_m^{(0)}$, and the REPN vectors $\Delta\tilde{\mathbf{y}}_m^{(0)}$ in (6.48). The hard decision estimates require the delay-time vectors to be transformed to the delay-Doppler domain and back using *two* N-IFFT operations, requiring $2N\log_2(N)$ complex multiplications per symbol vector. Hence, the overall complexity is $O(NM|\mathcal{L}|\mathrm{n_{iter}})$, where $\mathrm{n_{iter}}$ is the number of iterations.

[3]Divisions are counted as multiplications.

Algorithm 3: Reduced complexity MRC in delay-time domain.

Input: $\tilde{\mathbf{d}}_m, \tilde{\mathbf{x}}_m^{(0)}\ \forall\, m = 0, \ldots, M-1$
Input: $\tilde{\mathbf{H}} \in \mathbb{C}^{NM' \times NM'}, \tilde{\mathbf{y}}_m\ \forall\, m = 0, \ldots, M'-1$
for $m = 0 : M'-1$ **do**
  $\Delta\tilde{\mathbf{y}}_m^{(0)} = \tilde{\mathbf{y}}_m - \sum_{l\in\mathcal{L}} \tilde{\boldsymbol{\nu}}_{m,l} \circ \tilde{\mathbf{x}}_{m-l}^{(0)}$
end
for *i=1:max iterations* **do**
  $\Delta\tilde{\mathbf{y}}^{(i)} = \Delta\tilde{\mathbf{y}}^{(i-1)}$
  for $m = 0 : M-1$ **do**
    $\Delta\tilde{\mathbf{g}}_m^{(i)} = \sum_{l\in\mathcal{L}} \tilde{\boldsymbol{\nu}}_{m+l,l}^{*} \circ \Delta\tilde{\mathbf{y}}_{m+l}^{(i)}$
    $\tilde{\mathbf{c}}_m^{(i)} = \tilde{\mathbf{x}}_m^{(i-1)} + \Delta\tilde{\mathbf{g}}_m^{(i)} \oslash \tilde{\mathbf{d}}_m$
    $\tilde{\mathbf{x}}_m^{(i)} = \mathbf{F}_N^{\dagger} \cdot \mathcal{D}(\mathbf{F}_N \cdot \tilde{\mathbf{c}}_m^{(i)}) \quad (\text{or} \quad \tilde{\mathbf{x}}_m^{(i)} = \tilde{\mathbf{c}}_m^{(i)})$
    for $l \in \mathcal{L}$ **do**
      $\Delta\tilde{\mathbf{y}}_{m+l}^{(i)} \leftarrow \Delta\tilde{\mathbf{y}}_{m+l}^{(i)} - \tilde{\boldsymbol{\nu}}_{m+l,l} \circ (\tilde{\mathbf{x}}_m^{(i)} - \tilde{\mathbf{x}}_m^{(i-1)})$
    end
  end
  if $(||\Delta\tilde{\mathbf{y}}^{(i)}|| \geq ||\Delta\tilde{\mathbf{y}}^{(i-1)}||)$ **then EXIT**
end
Output: $\widehat{\mathbf{x}}_m = \mathcal{D}(\mathbf{F}_N \cdot \tilde{\mathbf{x}}_m)$

Initial step complexity

In Algorithm 3, the initial computations include generating all the entries of $\tilde{\mathbf{H}}$, which require computing $\tilde{\boldsymbol{\nu}}_{m,l}$ for $m = 0, \ldots, M'-1$ and $l \in \mathcal{L}$. Assuming the integer delay-Doppler channel parameters (h_i, l_i, k_i) are known for $i = 1, 2, \ldots, P$, where $P = \sum_{l\in\mathcal{L}} K_l$ and K_l denotes the number of integer Doppler taps in the same delay bin $l \in \mathcal{L}$, then $\tilde{\boldsymbol{\nu}}_{m,l}$ (N-IFFT of $\boldsymbol{\nu}_{m,l}$) can be computed in $\min\{NK_l, N\log_2(N)\}$ complex multiplications.

On the other hand, for the fractional Doppler case, the complexity of initial computations remains unaffected for the delay-time domain detector as $\tilde{\boldsymbol{\nu}}_{m,l}$ can be generated directly from the channel gains, delays, and Doppler shifts (h_i, ℓ_i, κ_i) of the P paths, using (2.13) and (4.98) with $M'NP$ complex multiplications.

6.5.5 Low complexity initial estimate

In Sections 6.5.1 and 6.5.3, we discussed the overall complexity of Algorithms 2 and 3, which scale linearly with the number of iterations. In both algorithms, we initially set $\widehat{\mathbf{x}}_m^{(0)} = \mathbf{0}_N$, for all m.

Next, we consider using a better initial estimate of the OTFS symbol vector, rather than $\widehat{\mathbf{x}}_m = \mathbf{0}_N$, to reduce the number of MRC iterations and improve detection and/or decoding convergence. Assuming the ideal pulse shaping waveform, we adopt a single-tap equalizer in the time-frequency domain to provide an improved low complexity initial estimate.

Following notations in Section 4.2, we let $\mathbf{H}_{\text{dd}} \in \mathbb{C}^{M\times N}$ be the delay-Doppler channel matrix for the ideal pulse shaping waveform with elements

$$\mathbf{H}_{\text{dd}}[m,n] = \begin{cases} \nu_l[\kappa], & \text{if } m = l, n = (\kappa)_N, \\ 0, & \text{otherwise,} \end{cases} \tag{6.50}$$

where $n = (\kappa)_N \in [0, N-1]$ and $\nu_l(\kappa)$ denotes the Doppler response of the l-th delay tap.

For the fractional Doppler case (i.e., κ is a real number), the ideal channel response can be written in terms of the Doppler spread vectors as

$$\mathbf{H}_{\text{dd}} = [\boldsymbol{\nu}_{0,0}, \boldsymbol{\nu}_{1,1}, \cdots, \boldsymbol{\nu}_{M-1,M-1}]^{\text{T}}$$

with $\boldsymbol{\nu}_{m,m} \in \mathbb{C}^{N\times 1}$, for $m = 0, \ldots, M-1$, where, from (4.80), we have

$$\boldsymbol{\nu}_{m,m}[k] = \frac{1}{N}\sum_{\kappa\in\mathcal{K}_l} \nu_l(\kappa)\zeta_N(\kappa - k)$$

for $k = 0, \ldots, N-1$, where $\mathcal{K}_l$ is the set of normalized Doppler shifts of all the paths with the same delay shift lT/M, and the periodic sinc function $\zeta_N(\cdot)$ caused by fractional Doppler shifts is given in (4.79).

The corresponding time-frequency channel response for the ideal pulse shaping waveform is obtained by an ISFFT operation on the delay-Doppler channel as

$$\begin{aligned}\mathbf{H}_{\text{tf}} &= \mathbf{F}_M \cdot \mathbf{H}_{\text{dd}} \cdot \mathbf{F}_N^{\dagger} \\ &= \mathbf{F}_M \cdot [\boldsymbol{\nu}_{0,0}, \boldsymbol{\nu}_{1,1}, \ldots, \boldsymbol{\nu}_{M-1,M-1}]^{\text{T}} \cdot \mathbf{F}_N^{\dagger} \\ &= \mathbf{F}_M \cdot [\tilde{\boldsymbol{\nu}}_{0,0}, \tilde{\boldsymbol{\nu}}_{1,1}, \ldots, \tilde{\boldsymbol{\nu}}_{M-1,M-1}]^{\text{T}}.\end{aligned} \tag{6.51}$$

Similarly, the received time-frequency samples can be obtained by the ISFFT operation on the received delay-Doppler domain samples as

$$\mathbf{Y}_{\text{tf}} = \mathbf{F}_M \cdot \mathbf{Y} \cdot \mathbf{F}_N^{\dagger} = \mathbf{F}_M \cdot [\tilde{\mathbf{y}}_0, \tilde{\mathbf{y}}_1, \cdots \tilde{\mathbf{y}}_{M-1}]^{\text{T}}. \tag{6.52}$$

In the case of ideal pulse shaping waveforms, since circular convolution of the channel and transmitted symbols in the delay-Doppler domain transforms to element-wise product in the time-frequency domain, as shown in Table 4.2, we can estimate the transmitted samples in the time-frequency domain by a single tap MMSE equalizer

$$\boxed{\widehat{\mathbf{X}}_{\text{tf}}[m,n] = \frac{\mathbf{H}_{\text{tf}}^*[m,n]\cdot\mathbf{Y}_{\text{tf}}[m,n]}{|\mathbf{H}_{\text{tf}}[m,n]|^2 + \sigma_w^2}} \tag{6.53}$$

for $m = 0, \ldots, M-1$ and $n = 0, \ldots, N-1$.

Let $[\tilde{\mathbf{x}}_0^{(0)}, \ldots, \tilde{\mathbf{x}}_m^{(0)}, \ldots, \tilde{\mathbf{x}}_{M-1}^{(0)}]^{\mathrm{T}}$ be the delay-time domain initial estimates of the OTFS symbol vectors, with $\tilde{\mathbf{x}}_m^{(0)} \in \mathbb{C}^{N\times 1}$, $m = 0, \ldots, M-1$. The estimates can be obtained as

$$[\tilde{\mathbf{x}}_0^{(0)}, \tilde{\mathbf{x}}_1^{(0)}, \cdots \tilde{\mathbf{x}}_{M-1}^{(0)}]^{\mathrm{T}} = \mathbf{F}_M^{\dagger} \cdot \widehat{\mathbf{X}}_{\mathrm{tf}}. \tag{6.54}$$

Complexity for initial estimate

Note that $\tilde{\boldsymbol{\nu}}_{m,l} = \mathbf{0}_N$ for $l \notin \mathcal{L}$ and hence the operation in (6.51) can be computed in $\min(NM|\mathcal{L}|, NM\log_2 M)$ complex multiplications. Since we already computed $\tilde{\boldsymbol{\nu}}_{m,l}$ and $\tilde{\mathbf{y}}$ is just a shuffled version of the received time domain samples, the overall number of computations involving (6.51), (6.52), (6.53), and (6.54) is upper bounded by $NM(|\mathcal{L}| + 2\log_2(M) + 3)$, which is comparable to the complexity per iteration of the delay-time implementation, i.e., $NM(2|\mathcal{L}| + 1)$, in Algorithm 3.

6.5.6 MRC detection for other OTFS variants

The MRC detection can be applied to other OTFS variants by choosing the appropriate submatrices $\mathbf{K}_{m,l}$ for each variant (see Chapter 4). The input–output relation in (6.33) can be modified to work with all OTFS variants as

$$\mathbf{y}_m = \sum_{l\in\mathcal{L}} \mathbf{K}_{m,l} \cdot \mathbf{x}_{[m-l]_M} + \mathbf{z}_m \tag{6.55}$$

for $m = 0, \ldots, M'-1$, $0 \le l \le l_{\max}$, where $M = M' - l_{\max}$ for ZP-OTFS and $M' = M$ for CP/RCP/RZP-OTFS.

Recall that, unlike ZP-OTFS, $\mathbf{K}_{m,l}$ in CP/RCP/RZP-OTFS are *nonzero* matrices for $m < l$ (see Table 4.2), leading to a slightly higher detection complexity than that of ZP-OTFS. Specifically, for RCP/RZP-OTFS, $\mathbf{K}_{m,l}$ for $m < l$ are not circulant matrices for $m < l$ and thus cannot be diagonalized in the delay-time domain, leading to additional detection complexity, when compared to CP/ZP-OTFS.

ZP-OTFS also converges faster compared to the other variants, and it has a better MRC estimate of $\mathbf{x}_0$ thanks to the ZP. Since the detectors work in the form of a DFE, the improved estimate of $\mathbf{x}_0$ leads to reduced error propagation as m is incremented. Other variants may achieve the same performance as that of ZP-OTFS after a larger number of iterations. Overall, the MRC algorithm is recommended for all OTFS variants, since the complexity of alternative solutions, such as LMMSE and MPA, is still much higher.

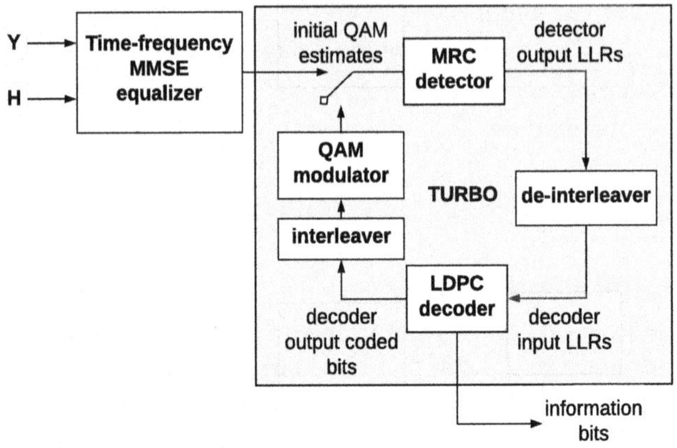

FIGURE 6.3 OTFS iterative rake turbo decoder operation.

6.6 Iterative rake turbo decoder

Next, we extend the uncoded OTFS system to the coded system, where we consider the low density parity check (LDPC) code as an error correction code (ECC) in the system.

At the transmitter, the information bits are encoded by an LDPC encoder, followed by a random interleaver. The interleaved coded bits are mapped to the QAM symbols, which are modulated via OTFS to generate the time domain signal. At the receiver, a joint MRC detection and decoding can be used in an iterative manner (i.e., an iterative rake turbo decoder), as shown in Fig. 6.3.

At first, the initial estimates of the QAM symbols provided by the proposed single tap MMSE equalizer in Section 6.5.5 are fed into the MRC detector, generating the bit-level log likelihood ratios (LLRs). The output LLRs are further processed by random deinterleaving and the LDPC decoding, generating hard decision coded bits. These coded bits are interleaved, QAM modulated, and fed back to the MRC detector as the updated input symbol vectors estimates. This completes one turbo iteration. The process repeats and the last turbo iteration outputs the LDPC decoded information bits.

From (6.42), the soft estimate of the delay-Doppler domain symbol vector $\mathbf{c}_m$ after MRC combining can be written as

$$\mathbf{c}_m = \mathbf{x}_m + \mathbf{e}_m, \qquad m = 0, \dots M-1, \tag{6.56}$$

where $\mathbf{x}_m$ is the transmitted symbol vector at delay index m and $\mathbf{e}_m$ denotes the normalized post-MRC NPI vector. We assume that $\mathbf{e}_m$ follows a zero mean Gaussian distribution with variance σ_m^2. This assumption becomes more accurate as the number of interfering terms in the MRC increases.

In the i-th iteration, we let $(\sigma_m^{(i)})^2$ be the variance of the NPI vector $\mathbf{e}_m^{(i)}$, and we let $L_{m,n,b}^{(i)}$ be the LLR of bit b of the n-th transmitted symbol in the estimated symbol vector $\mathbf{c}_m^{(i)}$, given by

$$
\begin{aligned}
L_{m,n,b}^{(i)} &= \log\left(\frac{Pr(b=0|\mathbf{c}_m^{(i)}[n])}{Pr(b=1|\mathbf{c}_m^{(i)}[n])}\right) \\
&= \log\left(\frac{\sum_{a\in\mathbb{A}_{b=0}}\exp(-|\mathbf{c}_m^{(i)}[n]-a|^2/(\sigma_m^{(i)})^2)}{\sum_{a'\in\mathbb{A}_{b=1}}\exp(-|\mathbf{c}_m^{(i)}[n]-a'|^2/(\sigma_m^{(i)})^2)}\right),
\end{aligned}
\tag{6.57}
$$

where $\mathbb{A}_{b=0}$ and $\mathbb{A}_{b=1}$ are subsets of the QAM alphabet $\mathbb{A}$, in which the b-th bit of the symbols is 0 and 1, respectively. The complexity of LLR computation can be reduced by using the max-log approximation as

$$
\tilde{L}_{m,n,b}^{(i)} = \frac{1}{(\sigma_m^{(i)})^2}\left(\min_{a\in\mathbb{A}_{b=0}}\left|\mathbf{c}_m^{(i)}[n]-a\right|^2 - \min_{a'\in\mathbb{A}_{b=1}}\left|\mathbf{c}_m^{(i)}[n]-a'\right|^2\right). \tag{6.58}
$$

To compute these LLRs in each iteration, an accurate estimate of $(\sigma_m^{(i)})^2$ is required, which is not straightforward to compute and requires knowledge of the correlation between all the estimated symbol and REPN vectors, which changes in each iteration. Since the entries of channel Doppler spread vectors $\boldsymbol{\nu}_{m,l}$ can be assumed to be independent and identically distributed normal random variables, the channel Doppler spread for different delay taps can be assumed to be uncorrelated, i.e., $\mathrm{E}[\boldsymbol{\nu}_{m,l}^{\dagger}\cdot\boldsymbol{\nu}_{m',l'}] = \delta_{m,m'}\delta_{l,l'}$.

Further, for the i-th iteration, we assume that the delay-Doppler REPN vectors $\Delta\mathbf{y}_m^{(i)}$ in the different delay branches are uncorrelated, i.e., $\mathrm{E}[(\Delta\mathbf{y}_m^{(i)})^{\dagger}\cdot\Delta\mathbf{y}_{m'}^{(i)}] = 0$ for $m \neq m'$, and follow Gaussian distributions. The covariance matrix of the delay-time REPN vector $\Delta\tilde{\mathbf{y}}_m^{(i)}$ in the i-th iteration is given by

$$
\boldsymbol{C}_m^{(i)}[j,k] = (\Delta\tilde{\mathbf{y}}_m^{(i)}[j]-\mathrm{E}(\Delta\tilde{\mathbf{y}}_m^{(i)}))(\Delta\tilde{\mathbf{y}}_m^{(i)}[k]-\mathrm{E}(\Delta\tilde{\mathbf{y}}_m^{(i)}))^{\dagger} \tag{6.59}
$$

for $j,k=0,\ldots,N-1$ and $\mathrm{E}(\Delta\tilde{\mathbf{y}}_m^{(i)}) = \frac{1}{N}\sum_{n=0}^{N-1}\Delta\tilde{\mathbf{y}}_m^{(i)}[n]$. Since Fourier transformation is a unitary transformation, the NPI variance remains the same in both domains, and we approximate this variance for the symbol vector soft estimate $\mathbf{c}_m^{(i)}$ in the i-th iteration as

$$
(\sigma_m^{(i)})^2 = \mathrm{Var}(\tilde{\mathbf{e}}_m^{(i)}) \approx \frac{1}{N}\sum_{l\in L}\eta_{m,l}\,\mathrm{tr}(\boldsymbol{C}_{m+l}^{(i)}), \tag{6.60}
$$

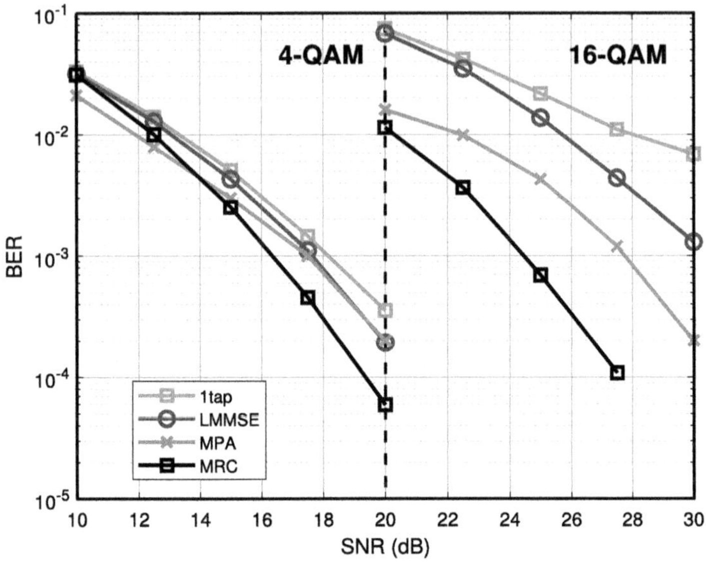

FIGURE 6.4 The BER performance of OTFS using 4-QAM and 16-QAM signaling and different detectors for a frame size of $N = M = 64$ (MP has 10 iterations for 4-QAM and 15 iterations for 16-QAM; MRC has 5 iterations for 4-QAM and 10 for 16-QAM).

where $\eta_{m,l} = ||\tilde{\mathbf{\nu}}_{m+l,l} \oslash \tilde{\mathbf{d}}_m||^2$ is the normalized post-MRC channel power in the different delay branches selected for combining. The LLR computation in (6.58) and NPI variance computation in (6.60) have a complexity of $NM \log_2 Q$ and $N^2 M|\mathcal{L}|$, respectively. The complexity of LDPC decoding is of the order $C_{\text{LDPC}} = O\left(NM \log_2 Q\right)$.

6.7 Illustrative results and discussion

In this section, we present simulation results for OTFS using various detection methods introduced in this chapter. Consider an OTFS frame with $N = M = 64$ (unless otherwise specified). The subcarrier spacing is $\Delta f = 15$ kHz and the carrier frequency is set to be 4 GHz. The standard channel models for simulations are specified in Section 2.5.1. For each SNR point, the BER is simulated over 10^4 OTFS frames.

Fig. 6.4 illustrates the bit error rate (BER) performance of the OTFS schemes with $M = N = 64$ using 4-QAM and 16-QAM signaling for an EVA channel model with a user equipment (UE) speed of 120 km/h. Different receiver detection schemes are considered in the simulation: a single-tap frequency domain equalization, LMMSE detection, an MP detection (10 and 15 iterations for 4- and 16-QAM signaling, respectively),

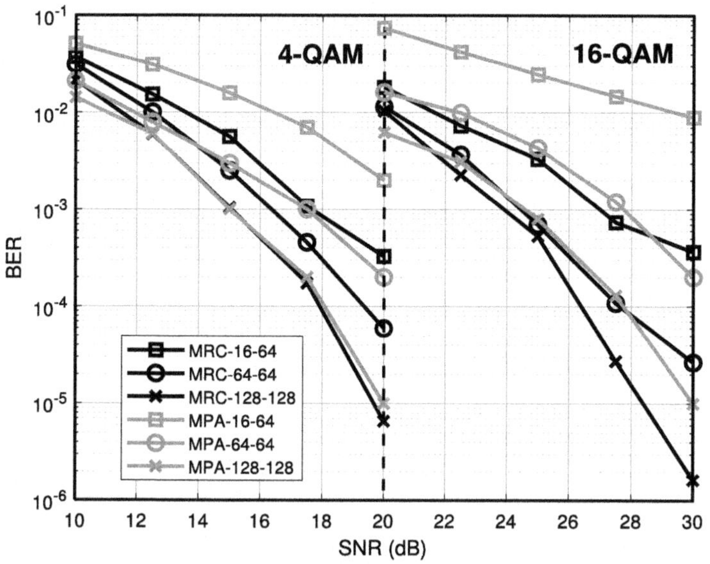

FIGURE 6.5 The BER performance of OTFS using MP and MRC detectors in the EVA channel model with different frame sizes (MP has 10 iterations for 4-QAM and 15 iterations for 16-QAM; MRC has 5 iterations for 4-QAM and 10 for 16-QAM).

and an MRC detection (5 and 10 iterations for 4- and 16-QAM signaling, respectively). We can observe that the MRC detection offers the best performance followed by MP and LMMSE, respectively. The lowest complexity single-tap equalizer offers the worst performance.

Fig. 6.5 shows the BER performance of the OTFS scheme with MRC and MP detections for different frame sizes with 4-QAM and 16-QAM signaling. It can be observed that the MP performance improves as the frame size increases, whereas MRC offers the best performance for all frame sizes. Further, the MRC detector offers good performance at much lower complexity as compared to the MP and LMMSE detection methods.

In Fig. 6.6, we simulate the BER performance of the OTFS schemes using the MRC detector under EPA, EVA, and ETU channel models with maximum Doppler spread for each path generated according to a maximum UE speed of 120 km/h. Both EVA and ETU model have 9 paths, whereas EPA is a 7-path channel model. The two frame sizes are chosen as $N = 64$ and $M = 64, 512$. The MRC detection uses 5 iterations. Increasing M improves the delay-resolution (T/M), thereby resolving the nearby path delays into different delay bins of the OTFS delay-Doppler grid. We observe from Fig. 6.6 that the OTFS BER performance improves with increasing frame size leading to more separable paths, which shows that OTFS can exploit channel diversity.

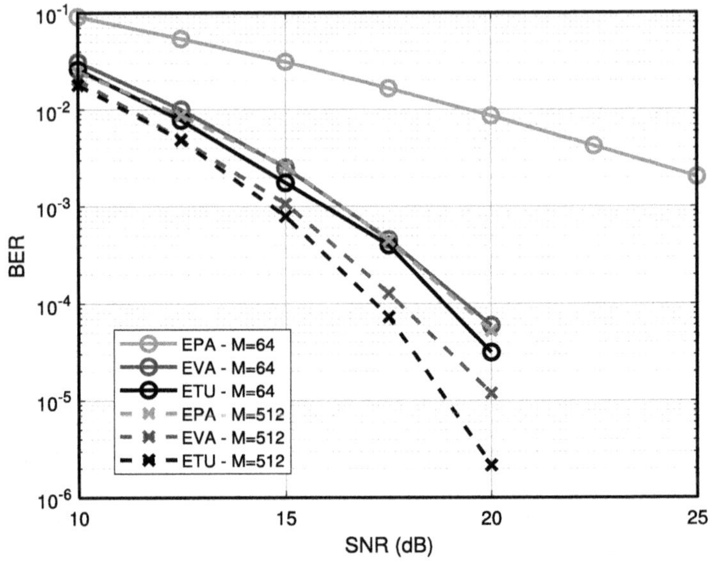

FIGURE 6.6 The BER performance of OTFS with 4-QAM signaling and the MRC detector (5 iterations) for EPA, EVA, and ETU channel models with max Doppler spread corresponding to a maximum speed of 120 km/h for $N = 64$ and $M = 64, 512$.

For the best error performance, the delay and Doppler resolution needs to be sufficient to resolve all the significant channel paths into distinct delay and Doppler bins, which is convenient for large frames. Consider the case in Fig. 6.6, where the frame sizes $M = 64$ and $M = 512$ correspond to a delay resolution of approximately 1000 ns and 130 ns. The EPA channel model has 7 paths with a maximum delay spread of 410 ns. For $M = 64$, none of the multipaths are separable along the delay domain, since all the paths fall within one delay bin of size 1000 ns. On the other hand, EVA and ETU channel models have 9 paths with a maximum delay spread of 2510 ns and 5000 ns, respectively, which can be separated into multiple delay bins at $M = 64$ and coherently combined at the receiver. This is demonstrated in Fig. 6.6, since OTFS has the worst performance for the EPA channel model. For $M = 512$, most of the energy in the EPA channel paths fall into 3 delay bins of size 130 ns, thereby allowing the OTFS detector to gain from delay diversity. For small sizes of M, the paths falling into the same delay bin can still be separated along the Doppler domain by increasing N, thereby taking advantage of the channel Doppler diversity.

To verify the OTFS performance in high mobility wireless channels, the BERs of OTFS at different UE speeds using 4-QAM signaling are plotted in Fig. 6.7, where the EVA channel model is considered. The OTFS system has $N = M = 64$ and uses the MRC detection with at most 10 iterations. We observe that the BER performance improves with increasing Doppler

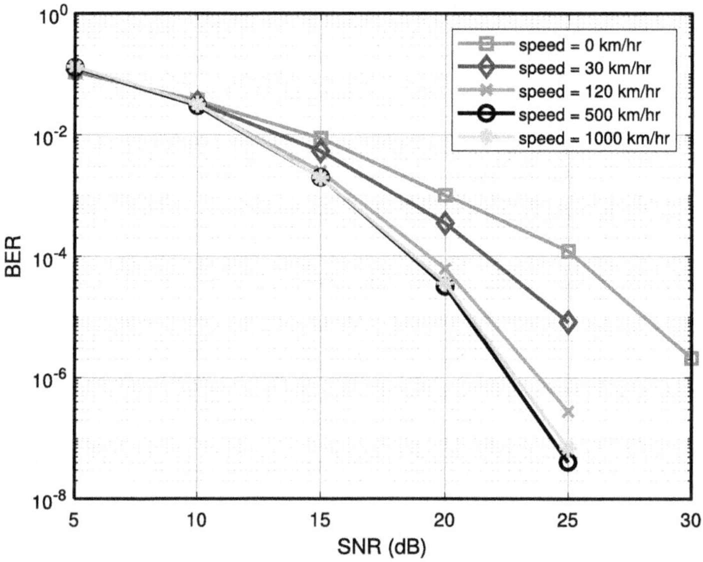

FIGURE 6.7 The BER performance of OTFS with 4-QAM signaling and the MRC detector for EVA at different UE speeds for $N = M = 64$.

spread. This is because when Doppler spread becomes greater than the Doppler resolution, those different paths falling within the same delay bin are separated into different Doppler bins. When the speed increases above 500 km/h, all paths are separated into different delay-Doppler bins. Hence, there is no significant increase in performance with increased Doppler spread.

In Fig. 6.8, we study the effect of multiple Doppler paths in each delay bin in OTFS with $N = M = 64$ and the MRC detection. The BER curve is plotted for 4-QAM signaling, for the synthetic wireless channel model described in Section 2.5.2 with $|\mathcal{L}| = 4$ delay taps. Each delay tap has multiple Doppler taps ranging from 1 to 4. It can be noted that the BER performance improves with multiple Doppler paths per delay, thereby showing that OTFS can exploit Doppler diversity. In the earlier plot for different channel models in Fig. 6.6, it was seen that the OTFS performance improves with an increased number of delay paths. The simulation results therefore indicate that OTFS can extract both delay and Doppler diversity in doubly selective channels.

6.8 Bibliographical notes

The MP detector for OTFS in Section 6.4 was presented in [1]. In-depth analysis, including the proof of convergence and methods to speed up the

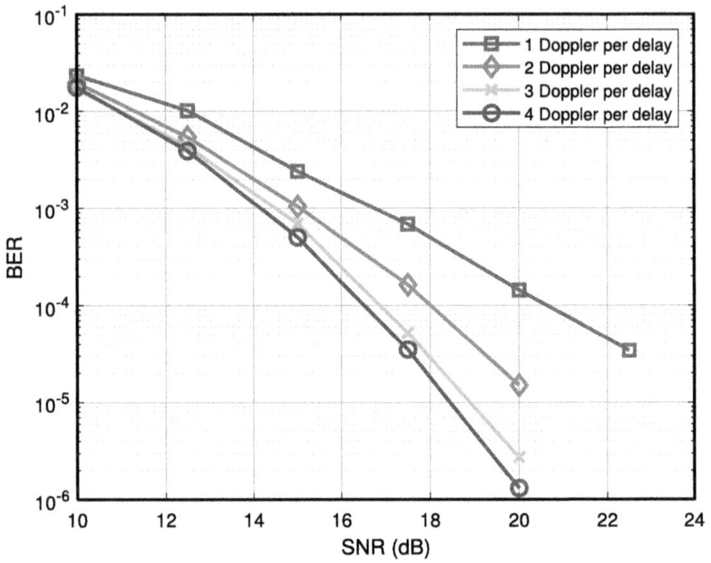

FIGURE 6.8 The BER performance of OTFS with 4-QAM signaling and the MRC detector for $N = 64$, $M = 64$ and different numbers of Doppler paths $|\mathcal{K}_l|$ per delay bin l using the synthetic wireless channel model described in Section 2.5.2.

detector convergence, can be found in [2,3]. Readers can refer to [4] for the basics of linear diversity combining techniques, including MRC. Analysis of performance of detection based on MRC of uncorrelated as well as correlated diversity branches can be found in [5]. The low complexity LMMSE detection and other equalization methods were studied in [6–14]. More iterative detection methods can be found in [14–25]. A real-time implementation of OTFS on an software-defined radio (SDR) platform using MP detection was done in [26].

References

[1] P. Raviteja, K.T. Phan, Y. Hong, E. Viterbo, Interference cancellation and iterative detection for orthogonal time frequency space modulation, IEEE Transactions on Wireless Communications 17 (10) (2018) 6501–6515, https://doi.org/10.1109/TWC.2018.2860011.

[2] T. Thaj, E. Viterbo, Low complexity iterative rake decision feedback equalizer for zero-padded OTFS systems, IEEE Transactions on Vehicular Technology 69 (12) (2020) 15606–15622, https://doi.org/10.1109/TVT.2020.3044276.

[3] T. Thaj, E. Viterbo, Low complexity iterative rake detector for orthogonal time frequency space modulation, in: 2020 IEEE Wireless Communications and Networking Conference (WCNC), 2020, pp. 1–6.

[4] D.G. Brennan, Linear diversity combining techniques, Proceedings of the IRE 47 (6) (1959) 1075–1102, https://doi.org/10.1109/JRPROC.1959.287136.

[5] X. Dong, N. Beaulieu, Optimal maximal ratio combining with correlated diversity branches, IEEE Communications Letters 6 (1) (2002) 22–24, https://doi.org/10.1109/4234.975486.
[6] G.D. Surabhi, A. Chockalingam, Low-complexity linear equalization for OTFS modulation, IEEE Communications Letters 24 (2) (2020) 330–334, https://doi.org/10.1109/LCOMM.2019.2956709.
[7] S. Tiwari, S.S. Das, V. Rangamgari, Low complexity LMMSE receiver for OTFS, IEEE Communications Letters 23 (12) (2019) 2205–2209, https://doi.org/10.1109/LCOMM.2019.2945564.
[8] S.S. Das, V. Rangamgari, S. Tiwari, S.C. Mondal, Time domain channel estimation and equalization of CP-OTFS under multiple fractional Dopplers and residual synchronization errors, IEEE Access 9 (2020) 10561–10576, https://doi.org/10.1109/ACCESS.2020.3046487.
[9] A. Pfadler, P. Jung, S. Stanczak, Mobility modes for pulse-shaped OTFS with linear equalizer, in: 2020 IEEE Global Communications Conference, 2020, pp. 1–6.
[10] T. Zou, W. Xu, H. Gao, Z. Bie, Z. Feng, Z. Ding, Low-complexity linear equalization for OTFS systems with rectangular waveforms, in: 2021 IEEE International Conference on Communications Workshops (ICC Workshops), 2021, pp. 1–6.
[11] C. Jin, Z. Bie, X. Lin, W. Xu, H. Gao, A simple two-stage equalizer for OTFS with rectangular windows, IEEE Communications Letters 25 (4) (2021) 1158–1162, https://doi.org/10.1109/LCOMM.2020.3043841.
[12] J. Feng, H. Ngo, M.F. Flanagan, M. Matthaiou, Performance analysis of OTFS-based uplink massive MIMO with ZF receivers, in: 2021 IEEE International Conference on Communications Workshops (ICC Workshops), 2021, pp. 1–6.
[13] Z. Ding, R. Schober, P. Fan, H.V. Poor, OTFS-NOMA: An efficient approach for exploiting heterogenous user mobility profiles, IEEE Transactions on Communications 67 (11) (2019) 7950–7965, https://doi.org/10.1109/TCOMM.2019.2932934.
[14] H. Qu, G. Liu, L. Zhang, S. Wen, M.A. Imran, Low-complexity symbol detection and interference cancellation for OTFS system, IEEE Transactions on Communications 69 (3) (2021) 1524–1537, https://doi.org/10.1109/TCOMM.2020.3043007.
[15] M. Ramachandran, A. Chockalingam, MIMO-OTFS in high-Doppler fading channels: signal detection and channel estimation, in: 2018 IEEE Global Communications Conference (GLOBECOM), 2018, pp. 1–6.
[16] F. Long, K. Niu, C. Dong, J. Lin, Low complexity iterative LMMSE-PIC equalizer for OTFS, in: 2019 IEEE International Conference on Communications (ICC), 2019, pp. 1–6.
[17] T. Zemen, M. Hofer, D. Löschenbrand, C. Pacher, Iterative detection for orthogonal precoding in doubly selective channels, in: 2018 IEEE 29th Annual International Symposium on Personal, Indoor and Mobile Radio Communications (PIMRC), 2018, pp. 1–7.
[18] W. Yuan, Z. Wei, J. Yuan, D.W.K. Ng, A simple variational Bayes detector for orthogonal time frequency space (OTFS) modulation, IEEE Transactions on Vehicular Technology 69 (7) (2020) 7976–7980, https://doi.org/10.1109/TVT.2020.2991443.
[19] R.M. Augustine, A. Chockalingam, Interleaved time-frequency multiple access using OTFS modulation, in: 2019 IEEE 90th Vehicular Technology Conference (VTC2019-Fall), 2019, pp. 1–5.
[20] L. Li, Y. Liang, P. Fan, Y. Guan, Low complexity detection algorithms for OTFS under rapidly time-varying channel, in: 2019 IEEE 89th Vehicular Technology Conference (VTC2019-Spring), 2019, pp. 1–6.
[21] Z. Yuan, F. Liu, W. Yuan, Q. Guo, Z. Wang, J. Yuan, Iterative detection for orthogonal time frequency space modulation with unitary approximate message passing, IEEE Transactions on Wireless Communications (2021) 1, https://doi.org/10.1109/TWC.2021.3097173.
[22] T. Thaj, E. Viterbo, Y. Hong, Orthogonal time sequency multiplexing modulation: analysis and low-complexity receiver design, IEEE Transactions on Wireless Communications 20 (12) (2021) 7842–7855, https://doi.org/10.1109/TWC.2021.3088479.

[23] K. Deka, A. Thomas, S. Sharma, OTFS-SCMA: a code-domain NOMA approach for orthogonal time frequency space modulation, IEEE Transactions on Communications 69 (8) (2021) 5043–5058, https://doi.org/10.1109/tcomm.2021.3075237.

[24] S. Li, W. Yuan, Z. Wei, J. Yuan, B. Bai, D.W.K. Ng, Y. Xie, Hybrid MAP and PIC detection for OTFS modulation, IEEE Transactions on Vehicular Technology 70 (7) (2021) 7193–7198, https://doi.org/10.1109/TVT.2021.3083181.

[25] H. Zhang, T. Zhang, A low-complexity message passing detector for OTFS modulation with probability clipping, IEEE Wireless Communications Letters 10 (6) (2021) 1271–1275, https://doi.org/10.1109/LWC.2021.3063904.

[26] T. Thaj, E. Viterbo, OTFS modem SDR implementation and experimental study of receiver impairment effects, in: 2019 IEEE International Conference on Communications Workshops (ICC Workshops), 2019, pp. 1–6.

CHAPTER

7

Channel estimation methods

OUTLINE

Chapter points

- Embedded pilot-aided delay-Doppler channel estimation.
- Embedded pilot-aided delay-time channel estimation.
- Real-time OTFS software-defined radio implementation.

Everyone sees what you appear to be, few experience what you really are.
Niccolò Macchiavelli

https://doi.org/10.1016/B978-0-32-385028-5.00015-3

7.1 Introduction

In this chapter, we focus on the input–output relation for OTFS pilot symbols for the purpose of channel estimation. Assuming integer delay and Doppler taps, we recall the delay-Doppler input–output relation in (4.123) only for the case $m \geq l_i$ as

$$\boxed{\mathbf{Y}[m,n] = \sum_{i=1}^{P} g_i z^{k_i(m-l_i)} \mathbf{X}[[m-l_i]_M, [n-k_i]_N],} \tag{7.1}$$

where $m = 0, \ldots, M-1$, $n = 0, \ldots, N-1$, $z = e^{\frac{j2\pi}{MN}}$, g_i, $i = 1, \ldots, P$, is the complex channel gain of the i-th path, with integer delay taps $l_i \leq l_{\max}$ (the largest channel delay tap) and integer Doppler taps k_i, and $\mathbf{X}, \mathbf{Y} \in \mathbb{C}^{M \times N}$ are the transmitted and received OTFS samples matrices.

Note that, in (7.1), we can ignore the case $m < l_i$ in (4.123), since we place the pilot symbol at the delay index

$$\begin{aligned} l_{\max} < m_p < M - l_{\max} \quad & \text{for CP/RCP/RZP-OTFS,} \\ 0 \leq m_p < M - l_{\max} \quad & \text{for ZP-OTFS.} \end{aligned} \tag{7.2}$$

Recall from Chapter 2 that in wideband systems, the actual channel delay shifts τ_i can be approximated to be integer multiples of the sampling period $1/M\Delta f$, i.e., $\tau_i = l_i/M\Delta f$, where $l_i \in \mathbb{Z}$. For large OTFS frames (large values of N), the actual Doppler shifts ν_i can also be approximated to be integer multiples of the Doppler resolution $1/NT$, i.e., $\nu_i = k_i/NT$, where $k_i \in \mathbb{Z}$. A large N results in an OTFS frame of long duration NT, which may increase the possibilities of channel parameters changing within the frame, leading to degraded channel estimation. Hence, in general, we consider $N < M$. However, for small values of N, the fractional Doppler effect is more prominent since it causes leakage of data symbols to more than P delay-Doppler grid locations. Thus, in such cases, it is pertinent to consider the effect of fractional Doppler in the input–output relation.

Eq. (7.1) describes the input–output relation for OTFS with only integer Doppler taps. For fractional Doppler shifts, we follow the notations in Chapters 2 and 4, where $\kappa_i \in \mathbb{R}$ represents the normalized Doppler shift. Then, the input–output relation for *fractional Doppler shifts* are

$$\boxed{\mathbf{Y}[m,n] = \sum_{i=1}^{P} g_i z^{\kappa_i(m-l_i)} \left(\sum_{k=0}^{N-1} \zeta_N(\kappa_i - k) \mathbf{X}[[m-l_i]_M, [n-k]_N] \right),} \tag{7.3}$$

where we recall from (4.79) the normalized periodic sinc function as

$$\zeta_N(x) = \frac{1}{N}\sum_{k=0}^{N-1} \mathrm{e}^{\frac{j2\pi xk}{N}} = \frac{1}{N}\frac{\sin(\pi x)}{\sin(\pi x/N)}\mathrm{e}^{\frac{j\pi x(N-1)}{N}}.$$

It can be observed from (7.3) that due to κ_i, each $\mathbf{Y}[m,n]$ is the aggregation of PN transmitted information symbols. Each fractional Doppler path causes the data symbols to interfere with all symbols along the Doppler axis, as demonstrated by the convolution between function $\zeta_N(\cdot)$ and the N symbols along the Doppler domain. However, since $\zeta_N(x)$ decays as x approaches $N/2$ or $-N/2$, only a subset of N data symbols need to be considered to closely approximate the operation in (7.3). Let $N_i < N$ be the number of data symbols that cause significant interference due to the fractional Doppler shift of the i-th path. Then the input–output relation in (7.3) can be approximated as

$$\boxed{\mathbf{Y}[m,n] = \sum_{i=1}^{P} g_i z^{\kappa_i(m-l_i)}\left(\sum_{k=-N_i/2}^{N_i/2-1} \zeta_N(\kappa_i - k)\mathbf{X}[[m-l_i]_M, [n-k]_N]\right).} \tag{7.4}$$

Based on the approximation, each received $\mathbf{Y}[m,n]$ is the aggregation of $\sum_{i=1}^{P} N_i$ information symbols. It is clear from (7.3) and (7.4) that the channel representation accuracy reduces due to $N_i < N$.

From the received symbols $\mathbf{Y}[m,n]$, if the channel parameters g_i, τ_i, and ν_i (and the corresponding taps l_i and κ_i) are known, we can employ the algorithms presented in Chapter 6 to detect the data symbols $\mathbf{X}[m,n]$. Hence, it is imperative to obtain channel state information using the following channel estimation methods.

7.2 Embedded pilot delay-Doppler channel estimation

7.2.1 The integer Doppler case

Let us consider the following system setting. At the transmitter, the OTFS frame consists of one pilot symbol, N_g guard symbols, and $MN - N_g - 1$ data symbols, as represented below.

- Let x_p denote the pilot symbol with pilot SNR: $\mathrm{SNR}_p = |x_p|^2/\sigma_w^2$, where σ_w^2 denotes the AWGN variance.
- Let $x_d[m,n]$ be the data symbols with data SNR: $\mathrm{SNR}_d = \mathrm{E}(|x_d|^2)/\sigma_w^2$, located at location $[m,n]$ in the delay-Doppler information grid.
- Let 0 represent the guard symbol.

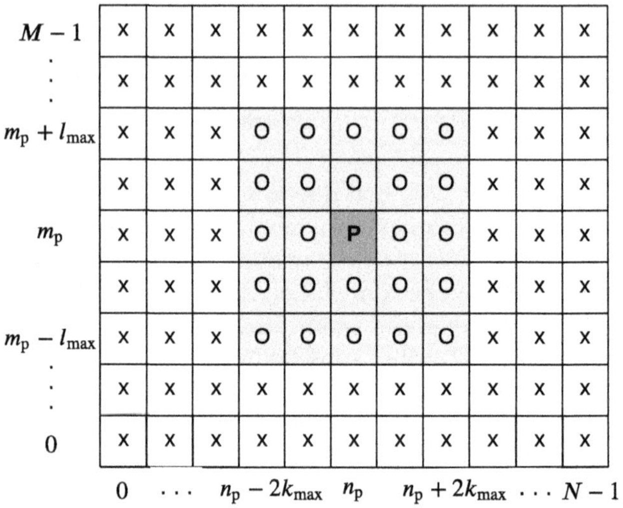

(a) Tx symbol arrangement (P: pilot; O: guard symbols; ×: data symbols)

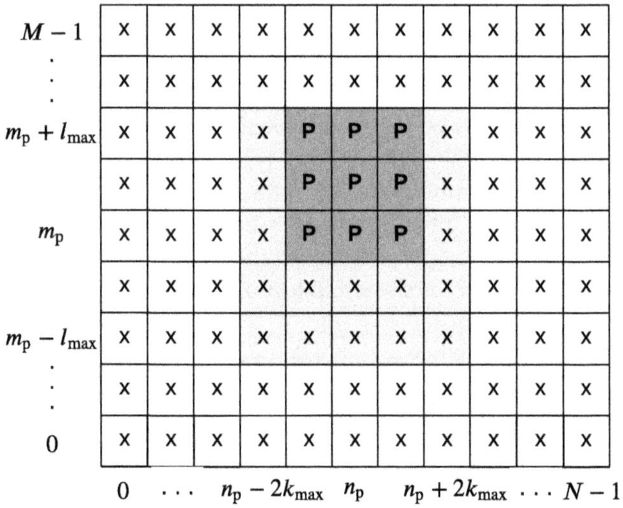

(b) Rx symbol pattern (P: channel estimation, ×: data detection)

FIGURE 7.1 (a) Transmitted pilot, guard, and data symbols. (b) Received symbols.

- Let $l_{\max}$ and $k_{\max}$ be the maximum delay and Doppler taps among the channel paths.

We arrange the pilot, guard, and data symbols in the delay-Doppler grid for transmission in an OTFS frame transmission (see Fig. 7.1a):

$$\mathbf{X}[m,n] = \begin{cases} x_p & m = m_p, n = n_p, \\ 0 & \begin{cases} m_p - l_{\max} \le m \le m_p + l_{\max}, \\ n_p - 2k_{\max} \le n \le n_p + 2k_{\max}, \end{cases} \\ x_d[m,n] & \text{otherwise,} \end{cases} \tag{7.5}$$

where we set (m_p, n_p) the location of the pilot symbol x_p, with m_p in (7.2) and $0 \le n_p - 2k_{\max} < n_p < n_p + 2k_{\max} \le N - 1$ for ease of representation.

With the help of N_g zero guard symbols, we arrange all symbols in this way to ensure there is no interference at the receiver between pilot and data symbols caused by channel delay and Doppler spread. Thus, we have $N_g = (2l_{\max} + 1)(4k_{\max} + 1) - 1$ guard symbols, with overhead

$$\frac{N_g + 1}{MN} = \frac{(2l_{\max} + 1)(4k_{\max} + 1)}{MN}$$

Typically, in LTE channels, the overhead for pilot and guard symbols is less than 1% of the data frame.

At the receiver, we adopt the received symbols $\mathbf{Y}[m,n]$, $m_p \le m \le m_p + l_{\max}$, $n_p - k_{\max} \le n \le n_p + k_{\max}$ for channel estimation, while the remaining received symbols $\mathbf{Y}[m,n]$ on the grid are used for data detection, as shown in Fig. 7.1b.

By replacing m and n by $m_p + l_i$ and $n_p + k_i$ in (7.1), the received pilot symbols for the i-th path, $i = 1, \ldots, P$, can be written in terms of the transmitted pilot symbol as

$$\begin{aligned} \mathbf{Y}[m_p + l_i, n_p + k_i] &= g_i z^{k_i m_p} \mathbf{X}[m_p, n_p] \\ &= g_i z^{k_i m_p} x_p. \end{aligned} \tag{7.6}$$

We aim to estimate channel parameters (g_i, l_i, k_i) for $i = 1, \ldots, P$, where *the number of propagation paths P is unknown*. We start with the received sample $\mathbf{Y}[m_p + l, n_p + k]$, $0 \le l \le l_{\max}$, $-k_{\max} \le k \le k_{\max}$, for channel estimation, and the remaining samples for data detection, as shown in Fig. 7.1. The estimated delay-Doppler channel gain is

$$\hat{g}[l,k] = \frac{\mathbf{Y}[m_p + l, n_p + k]}{x_p z^{k m_p}}. \tag{7.7}$$

However, in the presence of noise in (7.6), $\mathbf{Y}[m_p + l, n_p + k]$ may be mistaken for a channel path. Hence, we present below the threshold-based channel estimation scheme for path detection.

Let $b[l, k]$ denote whether a path with delay l and Doppler shift k exists as per the threshold criterion, i.e.,

$$b[l,k] = \begin{cases} 1, & |\mathbf{Y}[m_p + l, n_p + k]| \geq \mathcal{T}, \\ 0, & \text{otherwise}, \end{cases} \tag{7.8}$$

where the threshold $\mathcal{T}$ can be varied to alter the probabilities of miss detection and/or false alarm of path detection. The number of paths can be estimated as $\hat{P} = \sum_l \sum_k b[l, k]$. The delay index l_i and the Doppler shift index k_i, $i = 1, \ldots, \hat{P}$, correspond to the locations where $b[l, k] = 1$.

Then, the input–output relation in (7.1) can be rewritten in terms of the estimated channel parameters as

$$\mathbf{Y}[m,n] = \sum_{l=0}^{l_{\max}} \sum_{k=-k_{\max}}^{k_{\max}} \hat{g}[l,k] b[l,k] z^{k(m-l)} \mathbf{X}[[m-l]_M, [n-k]_N]. \tag{7.9}$$

The estimated Doppler response per delay tap defined in Chapter 4 can be written as

$$\hat{\nu}_l(k) = b[l,k]\hat{g}[l,k]. \tag{7.10}$$

The time-varying Doppler spread vector per delay tap $\hat{\boldsymbol{\nu}}_{m,l} \in \mathbb{C}^{N \times 1}$ can then be calculated for all the OTFS grid points as

$$\hat{\boldsymbol{\nu}}_{m,l}[[k]_N] = \hat{\nu}_l(k) z^{k(m-l)}. \tag{7.11}$$

The $MN \times MN$ delay-Doppler channel matrix $\mathbf{H}$ can be fully reconstructed with knowledge of $\hat{\nu}_l$ by applying the phase rotations $z^{k(m-l)}$ (see Chapter 4).

7.2.2 The fractional Doppler case

In the case of fractional Doppler, it can be seen from (7.3) that each propagation path spreads the pilot symbol into all the N Doppler shift indices (i.e., $k = 0, \ldots, N-1$) within a delay bin. Hence, by letting $m = m_p + l_i$ and $n = n_p + k$ in (7.3) for a given $l_i \in [0, l_{\max}]$ and $k \in [0, N-1]$, the received pilot symbols can be written in terms of the transmitted pilot symbols as

$$\begin{aligned} \mathbf{Y}[m_p + l_i, n_p + k] &= g_i z^{\kappa_i m_p} \zeta_N(\kappa_i - k) \mathbf{X}[m_p, n_p] \\ &= g_i z^{\kappa_i m_p} \zeta_N(\kappa_i - k) x_p, \end{aligned} \tag{7.12}$$

where $i = 1, \ldots, P$. Due to discretization of the delay-Doppler grid, the channel can be observed only at integer multiples of the delay and

Doppler resolution. Hence, the actual Doppler shifts (κ_i) are not known at the receiver, while, instead, one single path with fractional Doppler shift is observed as multiple paths with integer Doppler shifts adjacent to the fractional Doppler value.

Similar to the integer Doppler case, a threshold-based detection can be used to filter the false paths caused by the AWGN, as discussed below.

For simplification of notations, we omit subscript i in the following discussion. For a path with delay shift l and Doppler shift k, the discrete delay-Doppler channel is estimated as

$$\hat{g}[l,k] = \frac{\mathbf{Y}[m_p + l, n_p + k]}{x_p z^{km_p}}. \tag{7.13}$$

Similarly, $b[l,k]$ is defined based on threshold $\mathcal{T}$ as given in (7.8). However, unlike the integer Doppler case, in the absence of noise, a larger number of paths will be identified, i.e., $\sum_l \sum_k b[l,k] \geq P$, due to $\zeta_N(\kappa - k) \neq 0$ for $k = 0, \ldots, N-1$. This implies that we may end up estimating more paths than the actual number of propagation paths P due to fractional Doppler. However, some paths can be neglected as the periodic sinc function $\zeta(\kappa - k)$ decreases as $\kappa - k$ increases. To guarantee there is no interference between the received symbols for channel estimation and data detection, the guard symbols need to expand over a wider range over the Doppler axis, when compared to the integer Doppler case in (7.5). Therefore, we consider the following two cases to deal with fractional Doppler shifts.

- **Full guard symbols**
  Consider the following pilot placement:

$$\mathbf{X}[m,n] = \begin{cases} x_p & m = m_p, n = n_p, \\ 0 & m_p - l_{\max} \leq m \leq m_p + l_{\max}, \\ x_d[m,n] & \text{otherwise}. \end{cases} \tag{7.14}$$

  We refer to the case in (7.14) as *embedded pilot with full guard symbols*. We choose m_p as in (7.2) and $0 \leq n_p \leq N-1$. The number of guard symbols is $N_g = (2l_{\max}+1)N - 1$ and the overhead is $\frac{2l_{\max}+1}{M}$. Note that in typical LTE channels this may lead to about 8% overhead.
- **Reduced guard symbols**
  The full guard symbols offer better channel estimation at the cost of lower spectral efficiency by using more guard symbols and fewer data symbols. To improve the spectral efficiency, we can consider only the Doppler taps with significant energy. Now let us consider the following

pilot placement:

$$\mathbf{X}[m,n]=\begin{cases} x_p & l=m_p, k=n_p, \\ 0 & \begin{cases} m_p-l_{\max}\leq m\leq m_p+l_{\max}, \\ n_p-2k'_{\max}\leq n\leq n_p+2k'_{\max}, \end{cases} \\ x_d[m,n] & \text{otherwise.} \end{cases} \tag{7.15}$$

We refer to this case as *embedded pilot with reduced guard symbols*. We choose m_p as in (7.2) and $0 \leq n_p - 2k'_{\max} \leq n_p \leq n_p + 2k'_{\max} \leq N-1$. Here, $k'_{\max} = \lceil \kappa_{\max} \rceil + k_g$ denotes the maximum integer Doppler shift, where $\kappa_{\max}$ is the maximum fractional Doppler shift and k_g is the number of additional guard symbols needed on each side to collect all the significant pilot components in the received delay-Doppler grid for accurate channel estimation. Hence, the number of guard symbols is $N_g = (2l_{\max}+1)(4k'_{\max}+1)-1$ and the overhead is $\frac{(2l_{\max}+1)(4k'_{\max}+1)}{MN}$.

7.2.3 Effect of channel estimation on spectral efficiency

In Chapter 4, we described four OTFS variants and compared their design constraints in terms of normalized spectral efficiency (NSE),[1] transmit power, and channel sparsity in Table 4.4. It was shown that RZP/RCP-OTFS offer better NSE than ZP/CP-OTFS, thanks to a ZP/CP per frame rather than per block. In this section, we compare the effect of channel estimation on the NSE of these schemes since both pilot and guard symbols are needed for channel estimation.

Let $L_{\text{ZP}} = L_{\text{CP}} = l_{\max}$ be the length of ZP and CP, respectively. In Table 7.1, we compare the NSE of the OTFS variants without channel estimation vs. with embedded pilot channel estimation using reduced guard symbols $N_g = (2l_{\max}+1)(4k'_{\max}+1)-1$ and full guard symbols $(2l_{\max}+1)N-1$. From the table, we observe

- RZP/RCP-OTFS have the same NSEs in all cases;
- for low Doppler spread channels with reduced guard (small N_g), RZP/RCP-OTFS have the maximum NSE and ZP-OTFS has the second best solution, followed by CP-OTFS; and
- for high Doppler spread channels with full guard, ZP-OTFS offers the best NSE, since the ZP samples can be made part of the embedded pilot (see Fig. 7.2).

Moreover, for ZP-OTFS using full and reduced guard symbols, we can place the pilot symbol at $m_p = 0$ so that the pilot is spread by the channel into the ZP, thereby improving the NSE over CP-OTFS. Overall, the

[1]Here, normalized spectral efficiency is the ratio between the original spectral efficiency and $\log_2 Q$ for Q-QAM signaling.

ZP samples can prevent not only the interblock interference in the time domain, but also the interference between data and pilot samples.

As an example, we consider a frame of size $N = M = 64$ with 4-QAM signaling and a high mobility channel with $l_{\max} = 5$ samples (corresponding to the ETU channel). With full guard channel estimation, RZP/RCP-OTFS offer an NSE of 1.6542, CP-OTFS offers an NSE of 1.5362, and ZP-OTFS offers the highest NSE of 1.6812. The above benefits along with low transmit power and detection complexity make ZP-OTFS the best choice for high mobility channels.

TABLE 7.1 Normalized spectral efficiencies with/without channel estimation.

	no pilot + guard	**pilot + reduced guard**	**pilot + full guard**
CP	$\frac{M}{M+L_{CP}}$	$\frac{NM-(N_g+1)}{(M+L_{CP})N}$	$\frac{M-(2L_{CP}+1)}{M+L_{CP}}$
ZP	$\frac{M}{M+L_{ZP}}$	$\frac{NM-(N_g+1)\left(\frac{L_{ZP}+1}{2L_{ZP}+1}\right)}{(M+L_{ZP})N}$	$\frac{M-(L_{ZP}+1)}{M+L_{ZP}}$
RCP	$\frac{NM}{NM+L_{CP}}$	$\frac{NM-(N_g+1)}{NM+L_{CP}}$	$\frac{N(M-2L_{CP}-1)}{NM+L_{CP}}$
RZP	$\frac{NM}{NM+L_{ZP}}$	$\frac{NM-(N_g+1)}{NM+L_{ZP}}$	$\frac{N(M-2L_{ZP}-1)}{NM+L_{ZP}}$

7.3 Embedded pilot-aided delay-time domain channel estimation

In previous sections, we discussed the delay-Doppler channel estimation. In this section, we focus on ZP-OTFS in Fig. 4.13(c) and introduce a simpler delay-time channel estimation technique. As discussed in Chapter 4, for ZP-OTFS in Fig. 4.13(c), a ZP of length L_{ZP} is inserted in each block, so that the total frame size remains $M \times N$ samples. This is different from the one in Section 7.2.3 (see Fig. 4.13(b)), where a ZP is added to each block and the ZP removal leads to a scaling of the Doppler shifts by a factor $\gamma_g = \frac{M+L_{ZP}}{M}$. To avoid this scaling and simplify our exposition, we consider only the case in Fig. 4.13(c).

The delay-time domain channel estimation requires fewer channel parameters to represent a high mobility channel, when compared to the delay-Doppler domain channel estimation. As shown in Fig. 7.2, in time domain, due to the interleaved ZP containing pilot samples, the channel response for the l-th delay tap at time instants $\frac{(m_p+nM)}{M}T$ can be directly estimated, where $m_p + nM, n = 0, \ldots, N-1$, are the pilot locations in the time domain signal. Then a linear or spline interpolation is performed to reconstruct the time domain channel coefficients for the entire frame for each delay tap l. Further, we demonstrate that, for OTFS, linear interpolation is sufficient for accurate channel estimation at UE speeds under consideration (e.g., below 500 km/h).

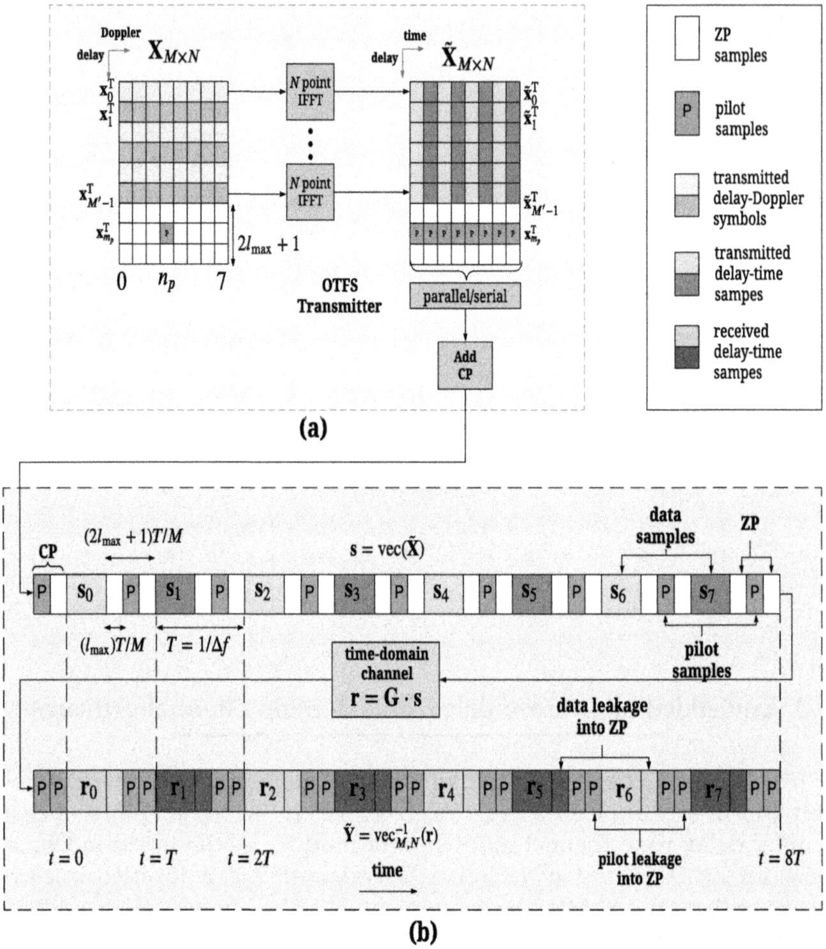

FIGURE 7.2 Pilot and data placement in ZP-OTFS ($N = 8$, $M = 9$) systems for time domain channel estimation. (a) ZP-OTFS transmitter. (b) Time domain operation.

7.3.1 Pilot placement

Fig. 7.2(a) illustrates the arrangement of pilot (denoted by "P" in the figure), guard, and data symbols in the delay-Doppler grid. We let $\mathbf{x}_m^T \in \mathbb{C}^{1\times N}$ be the m-th row vector in the OTFS matrix $\mathbf{X} \in \mathbb{C}^{M\times N}$, where $m = 0, \ldots, M-1$. We adopt the *same notations* as in Section 7.2.1, i.e., x_p and $x_d[m,n]$ denote the pilot and data symbols in locations (m_p, n_p) and (m, n), respectively, in the delay Doppler grid, and zero represents guard symbols. Thus, we have the following symbol arrangement:

$$\mathbf{x}_m^T[n] = \begin{cases} x_p & \text{if } m = m_p,\ n = n_p, \\ 0 & \text{if } n \neq n_p, 0 < |m - m_p| \leq l_{\max}, \\ x_d[m,n] & 0 \leq n \leq N-1, 0 \leq m \leq M'-1, \end{cases} \tag{7.16}$$

where $M' = M - 2l_{\max}$, $m_p = M' + l_{\max}$, $0 \leq n_p \leq N-1$, $n = 0, \ldots, N-1$, and $l_{\max}$ guard symbols are placed on each side of the pilot along the delay dimension to avoid interference at the receiver between data and pilot symbols due to delay and Doppler shifts. Hence, the number of zero guard symbols is $N_g = (2l_{\max}+1)N - 1$ and the overhead is $\frac{2l_{\max}+1}{M}$.

An N point IFFT is applied to each row vector $\mathbf{x}_m^T$. Hence, the pilot embedded row vector $\mathbf{x}_{m_p}^T$ is converted to

$$\tilde{\mathbf{x}}_{m_p}^T = \mathbf{x}_{m_p}^T \cdot \mathbf{F}_N^\dagger = x_p[\mathbf{F}_N^\dagger(n_p, 0), \cdots, \mathbf{F}_N^\dagger(n_p, N-1)]. \tag{7.17}$$

After the IFFT operations, the delay-time domain OTFS matrix is given by $\tilde{\mathbf{X}} = [\tilde{\mathbf{x}}_0, \ldots, \tilde{\mathbf{x}}_{M-1}]^T$, which is converted into a time domain vector $\mathbf{s} = \text{vec}(\tilde{\mathbf{X}})$ for transmission, as shown in Fig. 7.2(b). Then, in the time domain transmission, the interleaved pilot locations (green (mid gray in print version) boxes) allow parallel subsampling by a factor M in the entire OTFS frame. Since the location of the first pilot sample is at the sampling instant m_p, a CP is added to the start of the frame by copying the last $(l_{\max}+1)$ samples (which contain the pilot sample as well) of the OTFS frame. The pilot sample at location $m_p - M$ in the CP is needed to obtain the delay-time channel coefficients before the m_p-th sample by interpolation.

7.3.2 Delay-time channel estimation

The received signal in time domain $\mathbf{r} = \mathbf{Gs}$ is converted into delay-time domain $\tilde{\mathbf{Y}}^T = [\tilde{\mathbf{y}}_0, \ldots, \tilde{\mathbf{y}}_{M-1}]$, where $\mathbf{G}$ is the time domain channel matrix in (4.38). From the input–output relation in (4.96), replacing channel component $\tilde{\mathbf{v}}_{m,l}$ with $g^{\mathrm{s}}[l', m_p + l + nM]$ and then replacing m with $m_p + l$ for $l \in \mathcal{L}$ and summing over $l' \in \mathcal{L}$ yields the delay-time input–output relation

$$\tilde{\mathbf{y}}_{m_p+l}[n] = \sum_{l' \in \mathcal{L}} g^{\mathrm{s}}[l', m_p + l + nM] \tilde{\mathbf{x}}_{m_p+l-l'}[n]. \tag{7.18}$$

From (7.16), we know that $\tilde{\mathbf{x}}_{m_p+l-l'}[n] = 0$, when $0 < |l - l'| \leq l_{\max}$, due to the zero guard symbols. Then, from (7.18), the delay-time channel experienced by the pilot delay-time vector $\tilde{\mathbf{x}}_{m_p}$ can be simply estimated as

$$\hat{g}^{\mathrm{s}}[l, m_p + l + nM] = \frac{\tilde{\mathbf{y}}_{m_p+l}[n]}{\tilde{\mathbf{x}}_{m_p}[n]}, \quad \text{for } l \in \mathcal{L}. \tag{7.19}$$

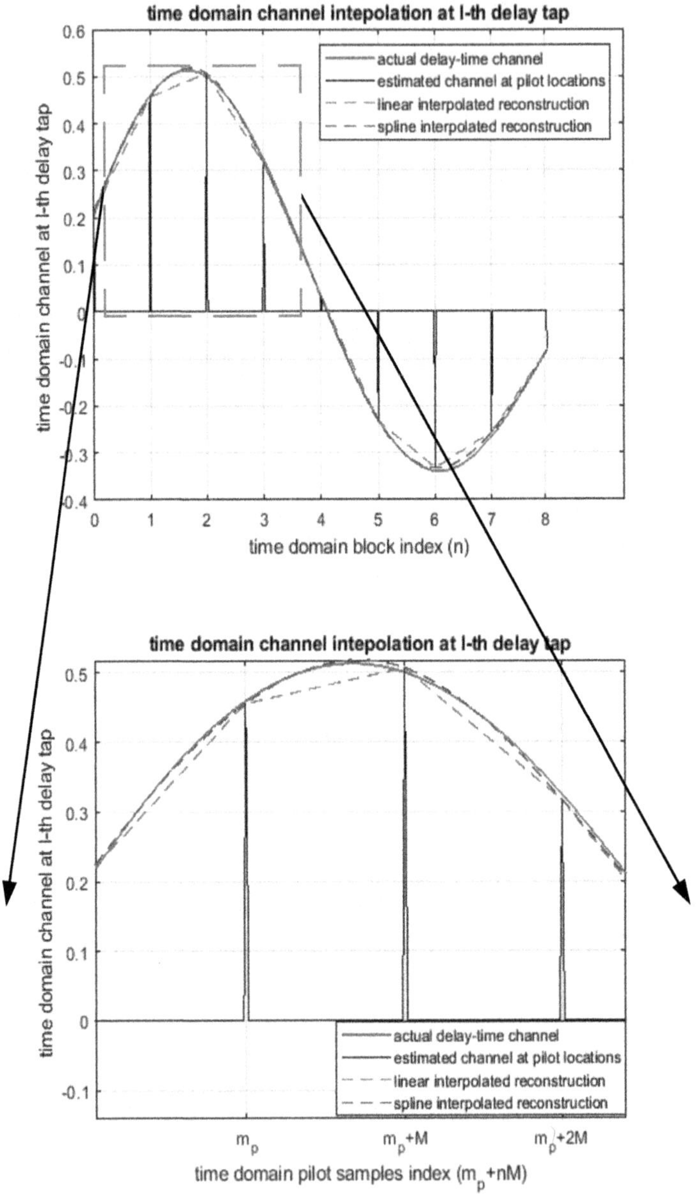

FIGURE 7.3 Reconstruction of the real part of the l-th delay tap channel from the estimated channel $\hat{g}^{s}[l, m_p + l + nM]$ using linear and spline interpolation for $N = 8$, $M = 64$ and UE speed = 500 km/h at $\text{SNR}_\text{d} = 20$ dB and $\beta = 0$ dB.

The estimated channel coefficients $\hat{g}^{s}[l, m_p + l + nM]$ can be considered as the subsampled delay-time channel at discrete pilot sample locations $m_p + nM$. The intermediate delay-time channel coefficients of the entire OTFS frame can be reconstructed by interpolating such samples.

Let f_s' be the subsampling frequency. According to the Nyquist sampling theorem, to accurately reconstruct a signal, f_s' needs to be at least twice the maximum frequency component of the signal, which is determined by the maximum Doppler shift $\nu_{\max}$ defined in (2.11). Due to subsampling by a factor M, we have

$$f_s' = f_s/M, \quad 2\nu_{\max} \le f_s' \le \Delta f.$$

Hence, the channel can be accurately reconstructed if

$$\nu_{\max} \le \frac{\Delta f}{2}.$$

This is a reasonable assumption for an underspread channel. The entire delay-time channel coefficients can be obtained by performing an interpolation of the estimated delay-time channel coefficients $\hat{g}^{s}[l, m_p + l + nM]$.

Fig. 7.3 shows an example of the real part of the time-varying channel at the l-th delay tap for a standard EVA channel model in Table 2.3 at a speed of 500 km/h. The transmitted pilot symbols can be viewed as a periodic delta function at intervals of T, one per each time domain block. The time variance of the delay-time channel coefficients is due to the different Doppler paths in the l-th delay tap. Since the effect of Doppler shifts can be modeled as the sum of sinusoidal functions, we use spline interpolation to reconstruct the time domain channel.

For low complexity channel estimation, we also consider linear interpolation. As shown in Fig. 7.3, linear interpolation fails to trace the delay-time response accurately when two consecutive interpolation points are very close to each other (see the zoomed-in section). Spline interpolation, however, can capture the channel variations better, when the delay-time channel is high Doppler spread.

To highlight the difference between the spline and linear interpolation methods, Fig. 7.4 shows the BER performance for 4-QAM for some extreme speeds of 500 km/h and 1000 km/h. It can be observed that linear interpolation performs similarly to spline interpolation for speeds no more than 500 km/h. This means that for underspread wireless channels, spline interpolation can be replaced with linear interpolation for low complexity channel estimation. Therefore, in the following, we consider only the linear interpolation method.

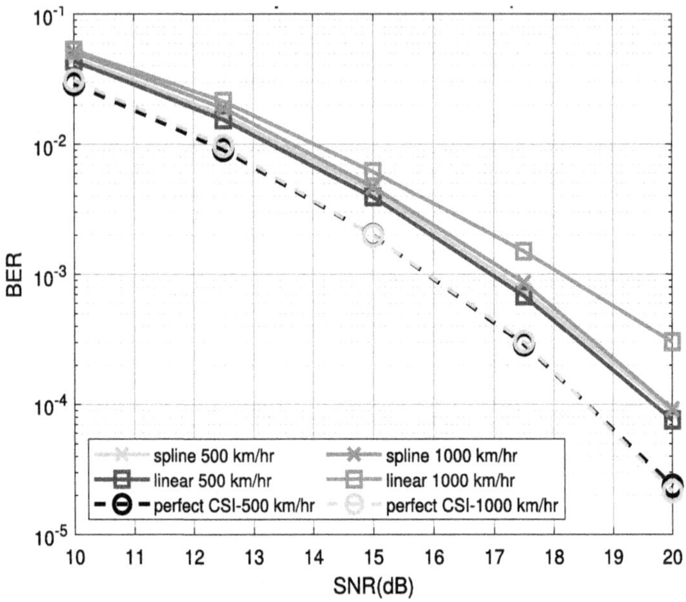

FIGURE 7.4 BER performance of 4-QAM OTFS using an iterative time domain detector for the EVA channel with speeds of 500 km/h and 1000 km/h.

7.3.3 Channel estimation complexity

From the initial step of channel estimation, we obtain the delay-time channel subsampling values $\hat{g}^{\mathrm{s}}[l, m_p+l+nM]$. Using linear interpolation, we define the n-th piece-wise slope of the estimated channel at l-th delay tap as

$$\alpha^{(n,l)} = \frac{\hat{g}^{\mathrm{s}}[l, m_p+l+(n+1)M] - \hat{g}^{\mathrm{s}}[l, m_p+l+nM]}{M} \tag{7.20}$$

for $-1 \leq n < N-1$. Then, the intermediate delay-time samples $\hat{g}^{\mathrm{s}}[l, m_p + l+u+nM]$ can be reconstructed as

$$\hat{g}^{\mathrm{s}}[l, m_p+l+u+nM] = \hat{g}^{\mathrm{s}}[l, m_p+l+nM] + \alpha^{(n,l)}u \tag{7.21}$$

for $0 < u < M-1$. It can be seen that (7.21) requires just one scalar multiplication with a complex number per sample (ignoring addition operations). Thanks to channel estimation, we already have NL out of NML delay-time channel coefficients. The initial operation in (7.21) then requires $2(N-1)ML$ real multiplications. The slope in (7.20) requires NL scaling operations (by $1/M$), which can be done using bit-shifting operations if M is a power of 2.

7.3.4 Extension to other OTFS variants

The delay-time channel estimation can be easily extended to other OTFS variants using embedded pilot schemes using full and reduced guard symbols. In the case of full guard symbols, the delay-time pilots have no interference from the data symbols and hence can be extracted from the time domain received signal, similar to ZP-OTFS. However, in the case of reduced guard symbols, the delay-time samples are a mix of data and pilot symbols. Therefore, the received OTFS signal is first converted to the delay-Doppler domain, where the pilot and data symbols are separated due to the presence of guard symbols. Since we know the Doppler positions of the received data symbols in the delay-Doppler grid, the related interference can be filtered out using simple masking along the Doppler domain. Finally, the received pilots are transformed back to the delay-time domain, where interpolation is carried out, similar to ZP-OTFS.

The delay-time interpolation method has another advantage: the receiver does not need to estimate the actual Doppler shifts of the channel paths. It was shown in Chapter 4 that the channel coefficients for the data symbols (consider only the $m \geq l$ case) in an OTFS frame have the same magnitude but vary in phase, depending on the location of the symbol. In the delay-Doppler channel estimation, the channel coefficients for the data symbols are computed by applying phase rotations to channel coefficients estimated from the pilot $\hat{g}[l,k]$ in (7.13). It can be noted from (7.3) that these phase rotations $z^{\kappa_i(m-l_i)}$ for $\mathbf{x}_m$ depend on the channel path's actual Doppler shift (κ_i), which is not straightforward to estimate in the case of fractional Doppler shifts.

As shown in Section 7.2.2, if the fractional Doppler shift κ_i is estimated as the nearest integer k_i (or a cluster of paths with integer Doppler shifts around κ_i), then the receiver applies the phase rotations wrongly as $z^{k_i(m-m_\text{p}-l_i)}$, instead of $z^{\kappa_i(m-m_\text{p}-l_i)}$. As a result, the error in phase rotation $z^{(\kappa_i-k_i)(m-m_\text{p}-l_i)}$ increases for m's farther away from pilot location m_p, leading to significant phase errors in the estimated channel coefficients. This is crucial for higher order modulation such as 64-QAM, where even small phase errors in channel coefficients can lead to large performance degradation.

The delay-time interpolation method on the other hand is free from this issue, since there is no need to directly estimate the delay-Doppler parameters. Moreover, it was shown in Chapter 6 that detection can be done with significantly lower complexity in the delay-time domain.

7.4 Real-time OTFS software-defined radio implementation

Like any other communications system, an OTFS modem is not free from receiver impairments such as DC offset and carrier frequency offset

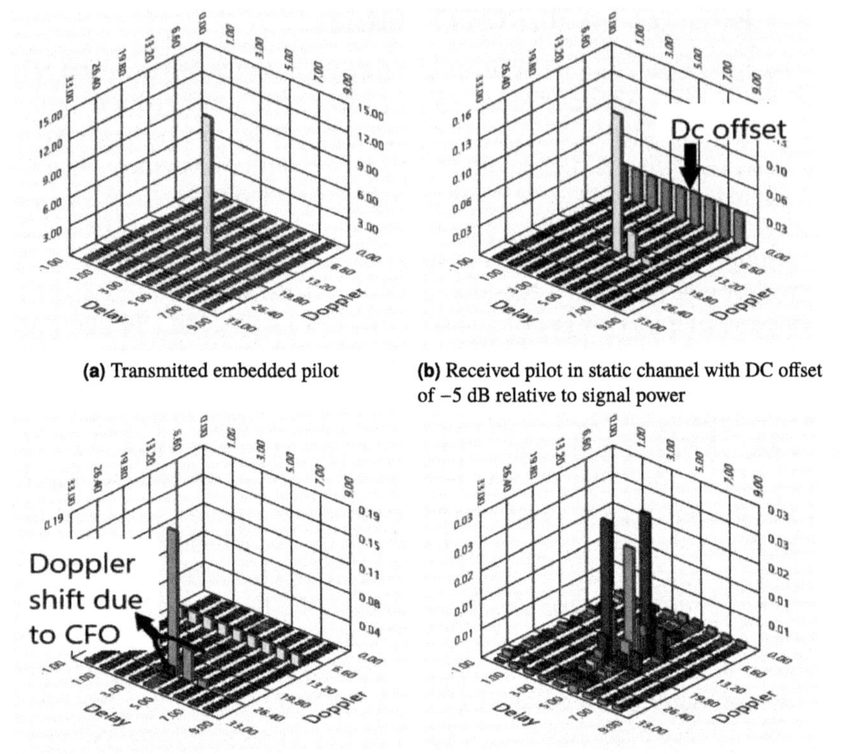

(a) Transmitted embedded pilot

(b) Received pilot in static channel with DC offset of −5 dB relative to signal power

(c) Received pilot with CFO of 150 kHz (five Doppler taps shift) in a static channel

(d) Received pilot with mobile channel emulator at the transmitter with three artificial paths

FIGURE 7.5 OTFS receiver impairment effects on the pilot (magnitude) in the indoor wireless channel (received SNR = 25 dB) (paths associated with the same Doppler shift [in the same row] are shaded with the same color).

(CFO), which affect channel estimation and detection. In this section, we study the effects of such impairments on the receiver performance via a real-time software-defined radio (SDR) implementation of an OTFS modem in a real indoor wireless channel. We compare the performance of OTFS with OFDM using the same SDR setup and environment settings for (i) the real indoor frequency selective (static) channel and (ii) the partially emulated mobility of a doubly selective channel.

Similar to the embedded pilot delay-Doppler channel estimation scheme discussed in Section 7.2.1, a pilot symbol is placed in the middle of a delay-Doppler grid of $M = N = 32$ in our experiments (see Fig. 7.5a). The pilot symbol is surrounded by zero guard symbols to avoid interference at the receiver, caused by delay and Doppler spread of data symbols. The number of zero guard symbols along delay and Doppler dimensions depends on the channel delay and Doppler shifts. Thanks to OTFS modulation, all the symbols in the frame undergo similar Doppler and delay shifts.

We conduct the OTFS experiments in an indoor wireless radio environment with $\tau_{\max} < 100$ ns. All the experiment parameters are given in Table 7.2. As observed in our channel measurements, we need only 4 delay taps to accommodate the delay spread with delay resolution 40 ns. Since there is no movement inside the room, the only Doppler shift is due to the carrier frequency offset. Once the channel information is extracted, the demodulated OTFS frame along with channel information is passed to the message passing detector, which recovers the QAM symbols. For our experiments, we adopt the threshold-based channel estimation as discussed in Section 7.2.1 with threshold $\mathcal{T} = \sqrt{3}\sigma_w$.

7.4.1 Effect of DC offset on channel estimation

Direct conversion receivers (DCRs), also known as zero intermediate frequency (zero-IF) or homodyne receivers, have become very popular, particularly in the realm of SDRs. There are many benefits in using DCRs such as the reduction in bulky off-chip front-end components, which leads to a higher level of integration and lower costs. But DCRs also come with some severe drawbacks such as DC offset and IQ imbalances.

DC offset manifests itself as a large spike in the center of the spectrum, which is due to the ADC being off by a half least significant bit (LSB) or the output of the low pass filters containing a DC bias. Another cause for DC offset could be at the mixer, where the local oscillator is performing self-mixing, i.e., the oscillator signal leaks back to the receiver front end and mixes with itself. Since DC offsets can have a negative impact on the receiver performance, it is important to estimate it.

In the received OTFS frame, as shown in Fig. 7.5b, the embedded pilot spreads according to the channel delay and Doppler shifts (yellow (light gray in print version) pulses). Further, it shows that the DC offset manifests itself as a constant signal in the zero Doppler shift region. This provides an opportunity to correct the DC offset by estimating and subtracting it from the zero Doppler shift row (first row in red (dark gray in print version)) of the OTFS frame. The DC offset can be estimated by taking an average of this zero Doppler shift row (first row in red (dark gray in print version)). To avoid the effect of DC offset on channel information, we reserve the first row of the frame for DC offset estimation.

7.4.2 Effect of carrier frequency offset on channel estimation

The mismatch between local oscillators at the transmitter and receiver introduces a CFO. In simulations, the CFO can be set to zero, which is not the case in reality. The IEEE 802.11a WLAN standard (based on OFDM waveform) specifies that the transmit center frequency error shall be a maximum of 20 ppm in both directions for the 5-GHz band. For the National Instruments universal software radio peripheral (NI USRP) SDR used in our experiments, a frequency accuracy of 2.5 ppm is being specified. Thus, for a 4-GHz carrier frequency we can expect a CFO range of ± 10 kHz.

An OFDM signal can be adversely affected by a small CFO, since it causes subcarriers to lose orthogonality. However, in OTFS, a CFO can be considered as a constant Doppler shift, experienced by a mobile receiver moving at a constant velocity in the same direction towards or away from the transmitter. In Fig. 7.5c, we show the effects of carrier shift on the pilot symbols, which demonstrates that both pilot symbols and DC offset undergo the same shift along the Doppler axis. This Doppler shift is equal to the CFO frequency and can be estimated and corrected using the message passing detector.

7.4.3 Experiment setup, results, and discussion

In the experiments, both transmitter and receiver terminals consist of two USRP-2943R SDRs, each connected to a host PC (see Fig. 7.6). The transmitter and receiver modem design and implementation are realized in LabView, running on the host PC. A LabView design or a program is called a virtual instrument (VI). Two VIs are needed for implementation, one for transmitter and the other for receiver. The ADC, the DAC, and the digital up/down conversion are realized via FPGA, while the rest of the digital signal processing including preamble detection and frame synchronization are done in LabView, using a combination of LabView Graphical Interface Blocks and C code. LabView can call functions written in C only

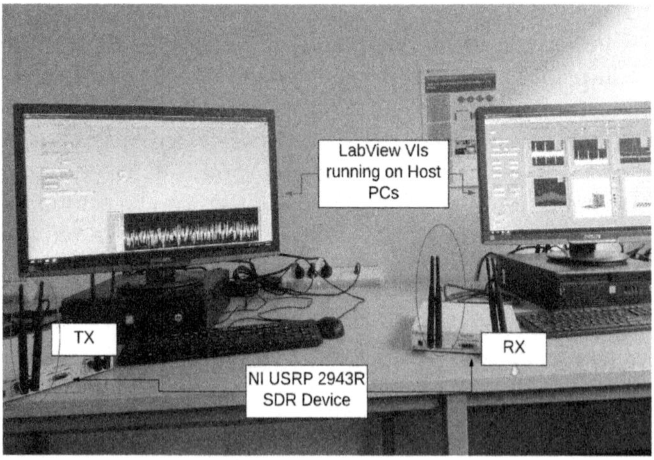

FIGURE 7.6 OTFS SDR modem setup.

as dynamic linked libraries (DLLs). The message passing detection algorithm is written as a function in a DLL file written in C. The carrier frequency, sampling rate, number of subcarriers, symbols, and modulation order can be set at run time of the VI. Transmitter and receiver gain can also be set at run time. The gain range of the USRP device is 0–31.5 dB.

For the experiments, we choose the modem parameters as in Table 7.2. Note that some of the experiment parameters in Table 7.2 have extreme values (such as for Δf) in order to resolve the delay-Doppler domain paths in an indoor channel (which in general has very few paths). We send OTFS frames continuously for each transmit gain configuration from 6.5 to 31.54 dB in steps of 5 dB, while keeping the receiver gain constant at 0 dB. The maximum transmit power (at a gain of 31.5 dB) at 4 GHz is in the range of 5 to 32 mW.

Fig. 7.7 shows the BER and frame error rate of the OTFS modem vs. transmit gain for 4-QAM and 16-QAM information symbols. The total transmit power of each OTFS frame is kept constant to normalize power for 4-QAM and 16-QAM. The measurements are averaged over 10,000 frames for each transmit gain configuration. We observe that the 4-QAM achieves better error performance compared to 16-QAM for the same transmit power.

TABLE 7.2 Experiment parameters.

Symbol	Parameter	Value
f_c	carrier frequency	4 GHz
M	number of subcarriers	32
N	number of symbols	32
Q	modulation alphabet size	4,16
T	symbol time	1.28 µs
Δf	subcarrier spacing	781.25 kHz
$1/M\Delta f$	delay resolution	40 ns
$1/NT$	Doppler resolution	24.4 kHz
$\nu_{\max}$	maximum emulator Doppler spread	400 kHz
d	Tx-Rx distance	1.5 meters

To emulate a mobile environment, we design and place a channel emulator block at the transmitter. This generates Doppler paths randomly from a uniform distribution with the specified maximum number of Doppler paths and maximum Doppler spread. We set 3 as the maximum number of Doppler taps and 400 kHz as the maximum Doppler spread ($\nu_{\max}$) (which equates to a maximum relative velocity of 30 km/h). The transmitter can then be imagined as a mobile transmitter traveling with a velocity that is related to the emulator parameters. The generated signal is then transmitted through a real-time wireless indoor channel that is frequency selective, hence simulating a doubly dispersive channel at the receiver, as shown in Fig. 7.5d.

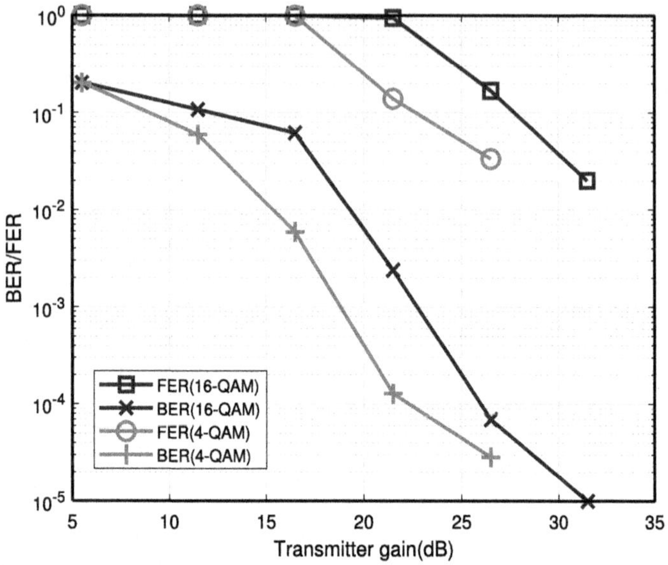

FIGURE 7.7 Bit and frame error rates vs. transmitter gain for 4-QAM and 16-QAM OTFS modulation.

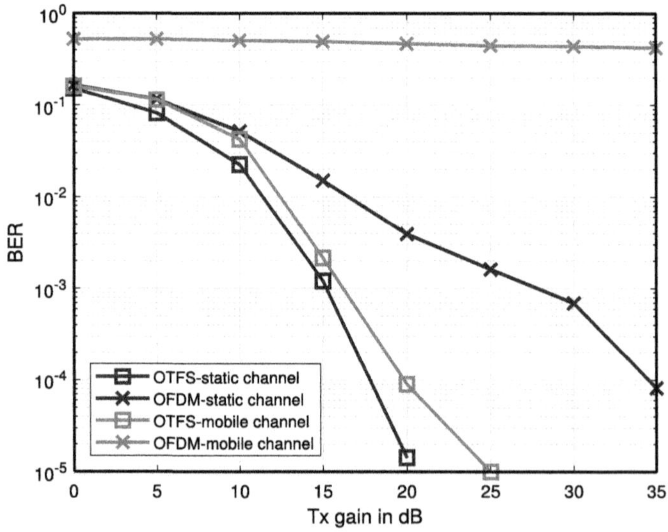

FIGURE 7.8 Bit error rates vs. transmitter gain for 4-QAM OTFS and OFDM modulation.

The performance of OTFS is compared with OFDM with a single tap equalizer using the same hardware setup for both time and frequency selective channels. For a fair comparison, an OTFS frame and an OFDM frame with the same number of pilot and data symbols are sent back to back, so that both frames undergo the same channel dispersion and hardware impairments. Fig. 7.8 illustrates the BERs of the schemes and demonstrates the superior performance of OTFS in both of the channel scenarios.

7.5 Bibliographical notes

The OTFS modulation was proposed by Hadani *et al.* in 2017 IEEE Wireless Communications and Networking Conference, San Francisco, CA, March 2017 [1] and the pilot-aided channel estimation with ideal pulse shaping waveform and integer Doppler shifts was discussed in [2]. The OTFS input–output relation and matrix representation adopted in this chapter were proposed in [3–5]. The embedded pilot-based channel estimations for OTFS presented in this chapter were proposed in [6–8] for the delay-Doppler and delay-time domains. Various channel estimation methods for OTFS systems with a single antenna, multiple antennas, a massive number of antennas, and multiple users can be found in [8–26]. More details regarding the real-time SDR implementation of an OTFS system with embedded pilot channel estimation was provided in [27].

References

[1] R. Hadani, S. Rakib, M. Tsatsanis, A. Monk, A.J. Goldsmith, A.F. Molisch, R. Calderbank, Orthogonal time frequency space modulation, in: 2017 IEEE Wireless Communications and Networking Conference (WCNC), 2017, pp. 1–6.

[2] R. Hadani, S. Rakib, Channel acquisition using orthogonal time frequency space modulated pilot signal, US patent US9444514B2, Sept. 13, 2016.

[3] P. Raviteja, K.T. Phan, Y. Hong, E. Viterbo, Interference cancellation and iterative detection for orthogonal time frequency space modulation, IEEE Transactions on Wireless Communications 17 (10) (2018) 6501–6515, https://doi.org/10.1109/TWC.2018.2860011.

[4] P. Raviteja, Y. Hong, E. Viterbo, E. Biglieri, Practical pulse-shaping waveforms for reduced-cyclic-prefix OTFS, IEEE Transactions on Vehicular Technology 68 (1) (2019) 957–961, https://doi.org/10.1109/TVT.2018.2878891.

[5] T. Thaj, E. Viterbo, Low complexity iterative rake decision feedback equalizer for zero-padded OTFS systems, IEEE Transactions on Vehicular Technology 69 (12) (2020) 15606–15622, https://doi.org/10.1109/TVT.2020.3044276.

[6] P. Raviteja, K. Phan, Y. Hong, E. Viterbo, Embedded delay-Doppler channel estimation for orthogonal time frequency space modulation, in: 2018 IEEE 88th Vehicular Technology Conference (VTC-Fall), 2018, pp. 1–6.

[7] P. Raviteja, K.T. Phan, Y. Hong, Embedded pilot-aided channel estimation for OTFS in delay-Doppler channels, IEEE Transactions on Vehicular Technology 68 (5) (2019) 4906–4917, https://doi.org/10.1109/TVT.2019.2906357.

[8] T. Thaj, E. Viterbo, Y. Hong, Orthogonal time sequency multiplexing modulation: analysis and low-complexity receiver design, IEEE Transactions on Wireless Communications 20 (12) (2021) 7842–7855, https://doi.org/10.1109/TWC.2021.3088479.

[9] K.R. Murali, A. Chockalingam, On OTFS modulation for high-Doppler fading channels, in: 2018 Information Theory and Applications Workshop (ITA), 2018, pp. 1–10.

[10] M. Ramachandran, A. Chockalingam, MIMO-OTFS in high-Doppler fading channels: signal detection and channel estimation, in: 2018 IEEE Global Communications Conference (GLOBECOM), 2018, pp. 1–6.

[11] W. Shen, L. Dai, J. An, P.Z. Fan, R.W. Heath, Channel estimation for orthogonal time frequency space (OTFS) massive MIMO, IEEE Transactions on Signal Processing 67 (16) (2019) 4204–4217, https://doi.org/10.1109/TSP.2019.2919411.

[12] W. Shen, L. Dai, S. Han, I.C. Lin, R.W. Heath, Channel estimation for orthogonal time frequency space (OTFS) massive MIMO, in: 2019 IEEE International Conference on Communications (ICC), 2019, pp. 1–6.

[13] Y. Hebron, S. Rakib, R. Hadani, M. Tsatsanis, C. Ambrose, J. Delfeld, R. Fanfelle, Channel acquisition using orthogonal time frequency space modulated pilot signal, US patent US10749651B2, Aug. 2020.

[14] O.K. Rasheed, G.D. Surabhi, A. Chockalingam, Sparse delay-Doppler channel estimation in rapidly time-varying channels for multiuser OTFS on the uplink, in: 2020 IEEE 91st Vehicular Technology Conference (VTC2020-Spring), 2020, pp. 1–6.

[15] Y. Liu, S. Zhang, F. Gao, J. Ma, X. Wang, Uplink-aided high mobility downlink channel estimation over massive MIMO-OTFS system, IEEE Journal on Selected Areas in Communications 38 (9) (2020) 1994–2009, https://doi.org/10.1109/JSAC.2020.3000884.

[16] F. Liu, Z. Yuan, Q. Guo, Z. Wang, P. Sun, Message passing based structured sparse signal recovery for estimation of OTFS channels with fractional Doppler shifts, IEEE Transactions on Wireless Communications 20 (12) (2021) 7773–7785, https://doi.org/10.1109/TWC.2021.3087501.

[17] L. Zhao, W.J. Gao, W. Guo, Sparse Bayesian learning of delay-Doppler channel for OTFS system, IEEE Communications Letters 24 (12) (2020) 2766–2769, https://doi.org/10.1109/LCOMM.2020.3021120.

[18] V.K. Singh, M.K. Flanagan, B. Cardiff, Maximum likelihood channel path detection and MMSE channel estimation in OTFS systems, in: 2020 IEEE 92nd Vehicular Technology Conference (VTC2020-Fall), 2020, pp. 1–6.

[19] S.S. Das, V. Rangamgari, S. Tiwari, S.C. Mondal, Time domain channel estimation and equalization of CP-OTFS under multiple fractional Dopplers and residual synchronization errors, IEEE Access 9 (2021) 10561–10576, https://doi.org/10.1109/ACCESS.2020.3046487.

[20] H. Qu, G. Liu, L. Zhang, M.A. Imran, S. Wen, Low-dimensional subspace estimation of continuous-Doppler-spread channel in OTFS systems, IEEE Transactions on Communications 69 (7) (2021) 4717–4731, https://doi.org/10.1109/TCOMM.2021.3072744.

[21] W. Yuan, S. Li, Z. Wei, J. Yuan, D.W.K. Ng, Data-aided channel estimation for OTFS systems with a superimposed pilot and data transmission scheme, IEEE Wireless Communications Letters 10 (9) (2021) 1954–1958, https://doi.org/10.1109/LWC.2021.3088836.

[22] S. Srivastava, R.K. Singh, A.K. Jagannatham, L. Hanzo, Bayesian learning aided sparse channel estimation for orthogonal time frequency space modulated systems, IEEE Transactions on Vehicular Technology 70 (8) (2021) 8343–8348, https://doi.org/10.1109/TVT.2021.3096432.

[23] S. Wang, J. Guo, X. Wang, W. Yuan, Z. Fei, Pilot design and optimization for OTFS modulation, IEEE Wireless Communications Letter 10 (8) (2021) 1742–1746, https://doi.org/10.1109/LWC.2021.3078527.

[24] L. Zhao, W.-J. Gao, W. Guo, Sparse Bayesian learning of delay-Doppler channel for OTFS system, IEEE Communications Letters 24 (12) (2020) 2766–2769, https://doi.org/10.1109/LCOMM.2020.3021120.

[25] C. Liu, S. Liu, Z. Mao, Y. Huang, H. Wang, Low-complexity parameter learning for OTFS modulation based automotive radar, in: ICASSP 2021–2021 IEEE International Conference on Acoustics, Speech and Signal Processing (ICASSP), 2021, pp. 8208–8212.

[26] S. Srivastava, R.K. Singh, A.K. Jagannatham, L. Hanzo, Bayesian learning aided sparse channel estimation for orthogonal time frequency space modulated systems, IEEE

Transactions on Vehicular Technology 70 (8) (2021) 8343–8348, https://doi.org/10.1109/TVT.2021.3096432.

[27] T. Thaj, E. Viterbo, OTFS modem SDR implementation and experimental study of receiver impairment effects, in: 2019 IEEE International Conference on Communications Workshops (ICC Workshops), 2019, pp. 1–6.

CHAPTER

8

MIMO and multiuser OTFS

OUTLINE

Chapter points

- MIMO-OTFS system model.
- MIMO-OTFS detection.

https://doi.org/10.1016/B978-0-32-385028-5.00016-5

- MIMO-OTFS channel estimation.
- Multiuser OTFS channel estimation.

We are drowning in information but starved for knowledge. **John Naisbitt**

8.1 Introduction

Multiple input multiple output (MIMO) is a wireless technology, where multiple transmit and receive antennas are simultaneously used for transmission and reception with an aim to increase channel capacity. MIMO provides high data rate communications by exploiting spatial diversity present in the channels between multiple transmit and receive antenna pairs.

Multiuser techniques are designed to provide high data rate and highly reliable communications to a large number of wireless terminals using a shared wireless medium. In a standard cellular system, multiuser uplink is formed when multiple terminals communicate to a base-station via a shared time/frequency resource, and the channel is known as the multiple access channel (MAC). A multiuser downlink is formed when a base-station broadcasts to multiple terminals, and the channel is known as the broadcast channel (BC).

In practice, OFDM has been widely used as a waveform in MIMO and multiuser systems. However, as discussed in the previous chapters for single input single output (SISO) cases, we have seen that OTFS can significantly outperform OFDM in high mobility scenarios, and perform similarly to OFDM in static or low mobility channels. In this chapter, we investigate OTFS for MIMO and multiuser communications, and present some detection and channel estimation methods for MIMO and multiuser systems.

8.2 System model for MIMO-OTFS

8.2.1 Transmitter and receiver

Consider a multiple input multiple output OTFS (MIMO-OTFS) system with n_T and n_R transmit and receiver antennas, as shown in Fig. 8.1. Let $\mathbf{X}^{(t)}$ and $\mathbf{Y}^{(r)}$ be the $M \times N$ delay-Doppler domain OTFS samples matrices, transmitted from the t-th antenna and received on the r-th antenna, for $t = 1, \ldots, n_\mathrm{T}$ and $r = 1, \ldots, n_\mathrm{R}$. In each link, the transmitted OTFS frame occupies a bandwidth of $M\Delta f$ and a duration of NT with $T = 1/\Delta f$, where Δf is the subcarrier spacing.

Assuming rectangular pulse shaping waveforms ($\mathbf{G}_{\mathrm{tx}} = \mathbf{G}_{\mathrm{rx}} = \mathbf{I}_M$), similar to (4.21) and (4.25) for the SISO case, the time domain samples transmitted from the t-th antenna and received on the r-th antenna are given by

$$\mathbf{s}^{(t)} = \text{vec}(\tilde{\mathbf{X}}^{(t)}) = \text{vec}(\mathbf{X}^{(t)} \cdot \mathbf{F}_N^{\dagger}) \in \mathbb{C}^{NM \times 1},$$
$$\mathbf{r}^{(r)} = \text{vec}(\tilde{\mathbf{Y}}^{(r)}) = \text{vec}(\mathbf{Y}^{(r)} \cdot \mathbf{F}_N^{\dagger}) \in \mathbb{C}^{NM \times 1} \quad (8.1)$$

for $t = 1, \ldots, n_{\text{T}}$ and $r = 1, \ldots, n_{\text{R}}$, where $\tilde{\mathbf{X}}^{(t)}, \tilde{\mathbf{Y}}^{(r)} \in \mathbb{C}^{M \times N}$ are the delay-time OTFS symbol matrices and $\mathbf{F}_N^{\dagger}$ is the N-point inverse Fourier transform.

In this chapter, we will consider ZP-OTFS for the MIMO case, where a ZP is added to each of the N blocks of an OTFS frame (see Fig. 4.13(b)), transmitted from each transmit antenna. The extension to the other variants of OTFS discussed in Chapter 4 is straightforward.

8.2.2 Channel

Let $P^{(r,t)}$ be the number of delay-Doppler paths in the channel between the r-th receive and t-th transmit antenna. For the i-th path in the channel between the r-th receive antenna and the t-th transmit antenna, where $i = 1, \ldots, P^{(r,t)}$, we let $g_i^{(r,t)}$ be the complex gain with delay shift $\tau_i^{(r,t)}$ and Doppler shift $\nu_i^{(r,t)}$. The delay-Doppler representation of the high mobility multipath channel is given by

$$h^{(r,t)}(\tau, \nu) = \sum_{i=1}^{P^{(r,t)}} g_i^{(r,t)} \delta(\tau - \tau_i^{(r,t)}) \delta(\nu - \nu_i^{(r,t)}) \quad (8.2)$$

for $t = 1, \ldots, n_{\text{T}}$ and $r = 1, \ldots, n_{\text{R}}$, and the corresponding delay-time channel between the (r, t) receive–transmit antenna pair can be written as

$$g^{(r,t)}(\tau, \theta) = \int_{\nu} h^{(r,t)}(\tau, \nu) e^{j2\pi\nu(\theta - \tau)} \, d\nu, \quad (8.3)$$

where θ is the continuous-time variable.

Let $\ell_i^{(r,t)} = \tau_i^{(r,t)} M \Delta f$ and $\kappa_i^{(r,t)} = \nu_i^{(r,t)} NT$ be the normalized delay and normalized Doppler shift associated with the i-th path for $i = 1, \ldots, P^{(r,t)}$ over the (r, t) antenna pair. The receiver samples the incoming signals at delay $\tau = \frac{l}{M\Delta f}$ and time $\theta = \frac{q}{M\Delta f}$, for $l, q \in \mathbb{Z}$, and thus the corresponding discrete-time equivalent channel $\bar{g}^{(r,t)}[l, q]$ can be written as

$$\bar{g}^{(r,t)}[l, q] = \sum_{i=1}^{P^{(r,t)}} g_i^{(r,t)} z^{(q-l)\kappa_i^{(r,t)}} \text{sinc}(l - \ell_i^{(r,t)}), \quad (8.4)$$

where $\mathrm{sinc}(x) = \sin(\pi x)/(\pi x)$. Assuming integer delay tap channels, i.e., $\ell_i^{(r,t)} = l_i^{(r,t)} \in \mathbb{Z}$, (8.4) reduces to

$$\bar{g}^{(r,t)}[l,q] = \sum_{i=1}^{P^{(r,t)}} g_i^{(r,t)} z^{(q-l)\kappa_i^{(r,t)}} \delta[l - l_i^{(r,t)}] \tag{8.5}$$

for $t = 1, \ldots, n_{\mathrm{T}}$ and $r = 1, \ldots, n_{\mathrm{R}}$, and we will consider *this case in the rest of the chapter.*

8.2.3 Input–output relation for MIMO-OTFS

8.2.3.1 Time domain

Using the discrete-time channel model in (8.5), the input–output relation between the (r,t) receive–transmit antenna pair can be written as

$$\mathbf{r}^{(r)}[q] = \sum_{t=1}^{n_{\mathrm{T}}} \sum_{l \in \mathcal{L}^{(r,t)}} \bar{g}^{(r,t)}[l,q]\mathbf{s}^{(t)}[q-l] + \mathbf{w}^{(r)}[q], \tag{8.6}$$

where $q = 0, \ldots, N(M + L_{\mathrm{ZP}}) - 1$ and L_{ZP} is the ZP length of at least $l_{\max}$ (the maximum channel delay tap among all channels). For ease of notations, we set $L_{\mathrm{ZP}} = l_{\max}$. In (8.6), $\mathbf{w}^{(r)}[q]$ denotes the AWGN at the r-th receive antenna, and $\mathcal{L}^{(r,t)} = \{l_i^{(r,t)}\}$ for $i = 1, \ldots, P^{(r,t)}$ denotes the set of distinct integer delay taps in the channel between the (r,t)-th antenna pair.

After ZP removal, we let $\mathbf{G}^{(r,t)} \in \mathbb{C}^{NM \times NM}$ be the time domain channel matrix between the (r,t) receive–transmit antenna pair with entries

$$\mathbf{G}^{(r,t)}[m + nM, m + nM - l] = \bar{g}^{(r,t)}[l, m + n(M + L_{\mathrm{ZP}})], \quad m \geq l, \tag{8.7}$$

for $l \in \mathcal{L}^{(r,t)}$ and *zero* otherwise, where, for ZP-OTFS, $\mathbf{G}^{(r,t)}$ is a lower triangular matrix, i.e., $\mathbf{G}^{(r,t)}[m + nM, m + nM - l] = 0$ for $m < l$ (see Fig. 4.19 in Chapter 4). Then, after ZP removal, the time domain input–output relation in (8.6) can be summarized in a simple vectorized form as

$$\boxed{\mathbf{r}^{(r)} = \sum_{t=1}^{n_{\mathrm{T}}} \mathbf{G}^{(r,t)}\mathbf{s}^{(t)} + \mathbf{w}^{(r)}} \tag{8.8}$$

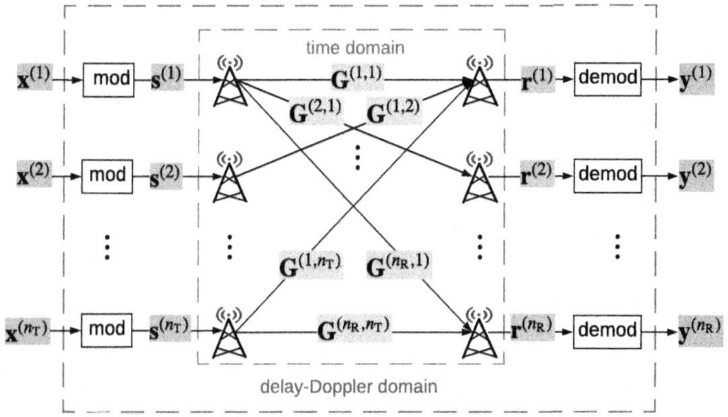

FIGURE 8.1 MIMO-OTFS system.

for $r = 1, \ldots, n_R$. The time domain MIMO input–output relation is then given as

$$\underbrace{\begin{bmatrix} \mathbf{r}^{(1)} \\ \mathbf{r}^{(2)} \\ \vdots \\ \mathbf{r}^{(n_R)} \end{bmatrix}}_{\mathbf{r}_{\text{MIMO}}} = \underbrace{\begin{bmatrix} \mathbf{G}^{(1,1)} & \mathbf{G}^{(1,2)} & \cdots & \mathbf{G}^{(1,n_T)} \\ \mathbf{G}^{(2,1)} & \mathbf{G}^{(2,2)} & \cdots & \mathbf{G}^{(2,n_T)} \\ \vdots & \ddots & \ddots & \vdots \\ \mathbf{G}^{(n_R,1)} & \mathbf{G}^{(n_R,2)} & \cdots & \mathbf{G}^{(n_R,n_T)} \end{bmatrix}}_{\mathcal{G}} \underbrace{\begin{bmatrix} \mathbf{s}^{(1)} \\ \mathbf{s}^{(2)} \\ \vdots \\ \mathbf{s}^{(n_T)} \end{bmatrix}}_{\mathbf{s}_{\text{MIMO}}} + \underbrace{\begin{bmatrix} \mathbf{w}^{(1)} \\ \mathbf{w}^{(2)} \\ \vdots \\ \mathbf{w}^{(n_R)} \end{bmatrix}}_{\mathbf{w}_{\text{MIMO}}}, \quad (8.9)$$

where $\mathbf{r}_{\text{MIMO}} \in \mathbb{C}^{NMn_R \times 1}$ and $\mathbf{s}_{\text{MIMO}} \in \mathbb{C}^{NMn_T \times 1}$ are the received and transmitted time domain samples, respectively, $\mathcal{G} \in \mathbb{C}^{NMn_R \times NMn_T}$ is the time domain MIMO channel matrix with submatrices $\mathbf{G}^{(r,t)}$, for all r, t, and $\mathbf{w}_{\text{MIMO}} \in \mathbb{C}^{NMn_R \times 1}$ is the AWGN vector.

8.2.3.2 *Delay-Doppler domain*

Let

$$\mathbf{x}^{(t)} = \text{vec}\left((\mathbf{X}^{(t)})^{\text{T}}\right), \quad \mathbf{y}^{(r)} = \text{vec}\left((\mathbf{Y}^{(r)})^{\text{T}}\right) \quad (8.10)$$

be the transmitted and received symbol vectors of size NM in the delay-Doppler domain. Recalling (4.32) in Section 4.4, the transmitted and received symbol vectors in time domain and delay-Doppler domain have the following relations:

$$\begin{aligned} \mathbf{s}^{(t)} &= \mathbf{P} \cdot (\mathbf{I}_M \otimes \mathbf{F}_N^{\dagger}) \cdot \mathbf{x}^{(t)}, \\ \mathbf{r}^{(r)} &= \mathbf{P} \cdot (\mathbf{I}_M \otimes \mathbf{F}_N^{\dagger}) \cdot \mathbf{y}^{(r)}, \end{aligned} \quad (8.11)$$

where $\mathbf{P} \in \mathbb{Z}^{NM \times NM}$ is the row-column interleaver matrix given in (4.33). Since $(\mathbf{P} \cdot (\mathbf{I}_M \otimes \mathbf{F}_N^\dagger))^\dagger = (\mathbf{I}_M \otimes \mathbf{F}_N)\mathbf{P}^T$, we can rewrite (8.11) as

$$\begin{aligned} \mathbf{x}^{(t)} &= (\mathbf{I}_\mathrm{M} \otimes \mathbf{F}_N)\mathbf{P}^T \cdot \mathbf{s}^{(t)}, \\ \mathbf{y}^{(r)} &= (\mathbf{I}_\mathrm{M} \otimes \mathbf{F}_N)\mathbf{P}^T \cdot \mathbf{r}^{(r)}. \end{aligned} \tag{8.12}$$

Following the derivation in Section 4.4.4, multiplying both sides of (8.8) with $(\mathbf{I}_M \otimes \mathbf{F}_N)\mathbf{P}^T$ yields the delay-Doppler input–output relation for the (r, t) antenna pair

$$\boxed{\mathbf{y}^{(r)} = \sum_{t=1}^{n_\mathrm{T}} \mathbf{H}^{(r,t)}\mathbf{x}^{(t)} + \mathbf{z}^{(r)},} \tag{8.13}$$

where

$$\mathbf{H}^{(r,t)} = (\mathbf{I}_M \otimes \mathbf{F}_N) \cdot (\mathbf{P}^\mathrm{T} \cdot \mathbf{G}^{(r,t)} \cdot \mathbf{P}) \cdot (\mathbf{I}_M \otimes \mathbf{F}_N^\dagger) \in \mathbb{C}^{NM \times NM}, \tag{8.14}$$

$$\mathbf{z}^{(r)} = (\mathbf{I}_M \otimes \mathbf{F}_N) \cdot (\mathbf{P}^\mathrm{T} \cdot \mathbf{w}^{(r)}) \in \mathbb{C}^{NM \times 1} \tag{8.15}$$

are the delay-Doppler channel matrix and the AWGN vector over the (r, t) antenna pair. As illustrated in Fig. 8.1, the delay-Doppler input–output relation for the entire MIMO system can be written as

$$\underbrace{\begin{bmatrix} \mathbf{y}^{(1)} \\ \mathbf{y}^{(2)} \\ \vdots \\ \mathbf{y}^{(n_\mathrm{R})} \end{bmatrix}}_{\mathbf{y}_\mathrm{MIMO}} = \underbrace{\begin{bmatrix} \mathbf{H}^{(1,1)} & \mathbf{H}^{(1,2)} & \cdots & \mathbf{H}^{(1,n_\mathrm{T})} \\ \mathbf{H}^{(2,1)} & \mathbf{H}^{(2,2)} & \cdots & \mathbf{H}^{(2,n_\mathrm{T})} \\ \vdots & \ddots & \ddots & \vdots \\ \mathbf{H}^{(n_\mathrm{R},1)} & \mathbf{H}^{(n_\mathrm{R},2)} & \cdots & \mathbf{H}^{(n_\mathrm{R},n_\mathrm{T})} \end{bmatrix}}_{\mathcal{H}} \underbrace{\begin{bmatrix} \mathbf{x}^{(1)} \\ \mathbf{x}^{(2)} \\ \vdots \\ \mathbf{x}^{(n_\mathrm{T})} \end{bmatrix}}_{\mathbf{x}_\mathrm{MIMO}} + \underbrace{\begin{bmatrix} \mathbf{z}^{(1)} \\ \mathbf{z}^{(2)} \\ \vdots \\ \mathbf{z}^{(n_\mathrm{R})} \end{bmatrix}}_{\mathbf{z}_\mathrm{MIMO}}, \tag{8.16}$$

where $\mathbf{y}_\mathrm{MIMO} \in \mathbb{C}^{NMn_\mathrm{R} \times 1}$ and $\mathbf{x}_\mathrm{MIMO} \in \mathbb{C}^{NMn_\mathrm{T} \times 1}$ are the received and transmitted delay-Doppler symbols, respectively, $\mathcal{H} \in \mathbb{C}^{NMn_\mathrm{R} \times NMn_\mathrm{T}}$ is the delay-Doppler MIMO channel matrix with submatrices $\mathbf{H}^{(r,t)}$, for all r, t, and $\mathbf{z}_\mathrm{MIMO} \in \mathbb{C}^{NMn_\mathrm{R} \times 1}$ is the AWGN vector.

It was shown in Chapter 4 that for rectangular pulse shaping waveforms, the $NM \times NM$ SISO channel matrix is comprised of circulant submatrices of size $N \times N$. Leveraging on this property, in order to derive the proposed detection method in Section 8.3.3, the transmit and received symbol vectors $\mathbf{x}^{(t)}, \mathbf{y}^{(r)} \in \mathbb{C}^{NM \times 1}$ are split into M subvectors of length N as

$$\mathbf{x}^{(t)} = \begin{bmatrix} \mathbf{x}_0^{(t)} \\ \vdots \\ \mathbf{x}_m^{(t)} \\ \vdots \\ \mathbf{x}_{M-1}^{(t)} \end{bmatrix}, \quad \mathbf{y}^{(r)} = \begin{bmatrix} \mathbf{y}_0^{(r)} \\ \vdots \\ \mathbf{y}_m^{(r)} \\ \vdots \\ \mathbf{y}_{M-1}^{(r)} \end{bmatrix}, \tag{8.17}$$

where $\mathbf{x}_m^{(t)}, \mathbf{y}_m^{(r)} \in \mathbb{C}^{N\times 1}$, $m = 0, \ldots, M-1$. Following the SISO notations in Chapter 4, the *delay-Doppler input–output relation* of the MIMO case in (8.16) after ZP removal can be written in a *subvectorized* form as

$$\boxed{\mathbf{y}_m^{(r)} = \sum_{t=1}^{n_\mathrm{T}} \sum_{l\in\mathcal{L}^{(r,t)}} \mathbf{K}_{m,l}^{(r,t)} \cdot \mathbf{x}_{m-l}^{(t)} + \mathbf{z}_m^{(r)},} \tag{8.18}$$

where $\mathbf{x}_{m-l}^{(t)} = \mathbf{0}$ for $m < l$, $\mathbf{K}_{m,l}^{(r,t)} \in \mathbb{C}^{N\times N}$ is the delay-Doppler channel matrix between the m-th received symbol vector of the r-th receive antenna and the $(m-l)$-th transmit symbol vector of the t-th transmit antenna.

8.2.3.3 Delay-time domain

Now we introduce the delay-time input–output relation for MIMO-OTFS, since some detection operations can be performed in this domain to save complexity. A tilde is superposed on the vectors to denote the corresponding delay-time domain variables related to those in the delay-Doppler domain by the N-point DFTs. The delay-time transmitted and received sample vectors $\tilde{\mathbf{x}}_m^{(t)}, \tilde{\mathbf{y}}_m^{(t)} \in \mathbb{C}^{N\times 1}$ are given by

$$\tilde{\mathbf{x}}_m^{(t)} = \mathbf{F}_N^\dagger \cdot \mathbf{x}_m^{(t)}, \quad \tilde{\mathbf{y}}_m^{(r)} = \mathbf{F}_N^\dagger \cdot \mathbf{y}_m^{(r)}. \tag{8.19}$$

Recall that since the delay-Doppler submatrices $\mathbf{K}_{m,l}$ are circulant matrices, (8.18) can be written in the form of element-wise multiplication in the delay-time domain as

$$\boxed{\tilde{\mathbf{y}}_m^{(r)}[n] = \sum_{t=1}^{n_\mathrm{T}} \sum_{l\in\mathcal{L}^{(r,t)}} \tilde{\boldsymbol{\nu}}_{m,l}^{(r,t)}[n]\tilde{\mathbf{x}}_{m-l}^{(t)}[n] + \tilde{\mathbf{w}}_m^{(r)}[n]} \tag{8.20}$$

for $n = 0, \ldots, N-1$, where $\tilde{\mathbf{x}}_{m-l}^{(t)} = \mathbf{0}$ for $m < l$, elements $\tilde{\mathbf{w}}_m^{(r)}[n]$, for all n, form the AWGN vector $\tilde{\mathbf{w}}_m^{(r)} \in \mathbb{C}^{N\times 1}$, and elements $\tilde{\boldsymbol{\nu}}_{m,l}^{(r,t)}[n]$, for all n, form the delay-time channel vector $\tilde{\boldsymbol{\nu}}_{m,l}^{(r,t)} \in \mathbb{C}^{N\times 1}$, which can be computed as

$$\tilde{\boldsymbol{\nu}}_{m,l}^{(r,t)} = \mathbf{F}_N^\dagger \cdot \boldsymbol{\nu}_{m,l}^{(r,t)}, \tag{8.21}$$

where $\boldsymbol{\nu}_{m,l}^{(r,t)} \in \mathbb{C}^{N\times 1}$ is the first column of the submatrix $\mathbf{K}_{m,l}^{(r,t)}$.

8.3 Detection methods

Now we are ready to present various detection methods for MIMO-OTFS. Let $S^{(r,t)}$ be the number of discrete delay-Doppler paths in the channel between the (r,t) antenna pair. Note that, assuming integer delay-Doppler paths in this chapter, we have $S^{(r,t)} = P^{(r,t)}$.

Since each row of the submatrix $\mathbf{H}^{(r,t)}$ has $S^{(r,t)}$ nonzero elements, it has a total of $NMS^{(r,t)}$ nonzero elements. Consequently, the entire MIMO channel matrix $\mathcal{H}$ has $\sum_r \sum_t NMS^{(r,t)}$ nonzero elements.

To simplify notations, we assume that there are S delay-Doppler paths on average in the channel between each antenna pair so that we have $NMSn_{\mathrm{R}}n_{\mathrm{T}}$ nonzero elements in $\mathcal{H}$. Similarly, assuming $L = |\mathcal{L}|^{(r,t)}$, for all (r,t) pairs, is the number of distinct delay taps on average in the channel between any antenna pair, the total number of nonzero elements in the time domain channel matrix $\mathcal{G}$ is $NMLn_{\mathrm{R}}n_{\mathrm{T}}$. Since $L \leq S$ as Doppler paths cannot be distinguished in time domain, $\mathcal{G}$ is sparser than $\mathcal{H}$. The difference in sparsity can be taken advantage of in reducing the complexity of detection and channel estimation.

8.3.1 Linear minimum mean-square error detector

Similar to the detection method for SISO-OTFS in Section 6.3, the LMMSE estimate for MIMO-OTFS in (8.16) in the delay-Doppler domain is given as

$$\boxed{\hat{\mathbf{x}}_{\mathrm{MIMO}} = (\mathcal{H}^{\dagger} \cdot \mathcal{H} + \sigma_z^2 \mathbf{I}_{MNn_{\mathrm{T}}})^{-1} \cdot \mathcal{H}^{\dagger} \cdot \mathbf{y}_{\mathrm{MIMO}},} \tag{8.22}$$

where σ_z^2 is the AWGN power. Solving (8.22) requires a large matrix inversion, where the matrix size is $NMn_{\mathrm{T}} \times NMn_{\mathrm{T}}$. This could incur a very high complexity of $O((NMn_{\mathrm{T}})^3)$. However, the sparsity of $\mathcal{H}$ can be utilized to reduce the complexity of the LMMSE detection.

Since the time domain channel matrix $\mathcal{G}$ is sparser than $\mathcal{H}$, the complexity may be reduced by performing the LMMSE detection in the time domain and then transforming the time domain estimates to the delay-Doppler domain to get back the information symbols. The time domain estimate of (8.9) is given as

$$\boxed{\hat{\mathbf{s}}_{\mathrm{MIMO}} = (\mathcal{G}^{\dagger} \cdot \mathcal{G} + \sigma_w^2 \mathbf{I}_{MNn_{\mathrm{T}}})^{-1} \cdot \mathcal{G}^{\dagger} \cdot \mathbf{r}_{\mathrm{MIMO}},} \tag{8.23}$$

where $\hat{\mathbf{s}}_{\mathrm{MIMO}} = \{\hat{\mathbf{s}}^{(t)}\}_{t=1}^{n_T}$. The estimated delay-Doppler information symbols transmitted by the t-th antenna can be obtained as

$$\hat{\mathbf{x}}^{(t)} = (\mathbf{I}_{\mathrm{M}} \otimes \mathbf{F}_N^{\dagger}) \cdot \mathbf{P}^{\mathrm{T}} \cdot \hat{\mathbf{s}}^{(t)}. \tag{8.24}$$

8.3.2 Message passing detector

Similar to the detection of SISO-OTFS in Section 6.4, the message passing (MP) can be applied to MIMO-OTFS as well. Consider the vectorized input–output relation in (8.16), where $\mathbf{y}_{\text{MIMO}}$, $\mathbf{z}_{\text{MIMO}}$ are $NMn_R \times 1$ complex vectors with elements $y_{\text{MIMO}}[d]$ and $z_{\text{MIMO}}[d]$, $\mathcal{H}$ is the $NMn_R \times NMn_T$ complex matrix with elements $\mathcal{H}[d,c]$, $\mathbf{x}_{\text{MIMO}}$ is an $NMn_T \times 1$ complex vector with elements $x_{\text{MIMO}}[c] \in \mathbb{A}$, where $d \in [0, NMn_{\text{R}} - 1]$, $c \in [0, NMn_{\text{T}} - 1]$, and $\mathbb{A} = \{a_1, \ldots, a_Q\}$ represents a modulation alphabet of size Q.

Let $\mathcal{I}(d)$ and $\mathcal{J}(c)$ denote the sets of indexes with nonzero elements in the d-th row and the c-th column, respectively. Then $|\mathcal{I}(d)| = \sum_{r=1}^{n_{\text{R}}} S^{(r,t)}$ for all rows and $|\mathcal{J}(c)| = \sum_{t=1}^{n_{\text{T}}} S^{(r,t)}$ for all columns.

From (8.16), the joint maximum a posteriori probability (MAP) detection rule for estimating the transmitted signals is given by

$$\hat{\mathbf{x}} = \arg \max_{\mathbf{x} \in \mathbb{A}^{NMn_{\text{T}} \times 1}} \Pr\left(\mathbf{x}_{\text{MIMO}} \,\middle|\, \mathbf{y}_{\text{MIMO}}, \mathcal{H}\right),$$

which has an exponential complexity in NMn_{T}, i.e., Q^{NMn_T}. Since the joint MAP detection can be intractable for practical values of N, M, and n_{R}, we consider the symbol-by-symbol MAP detection rule for $c = 0, \ldots, NMn_{\text{T}} - 1$,

$$\begin{aligned}
\hat{x}[c] &= \arg\max_{a_j \in \mathbb{A}} \Pr\left(x_{\text{MIMO}}[c] = a_j \,\middle|\, \mathbf{y}_{\text{MIMO}}, \mathcal{H}\right) \\
&= \arg\max_{a_j \in \mathbb{A}} \frac{1}{Q} \Pr\left(\mathbf{y}_{\text{MIMO}} \,\middle|\, x_{\text{MIMO}}[c] = a_j, \mathcal{H}\right) \qquad (8.25) \\
&\approx \arg\max_{a_j \in \mathbb{A}} \prod_{d \in \mathcal{J}_c} \Pr\left(y_{\text{MIMO}}[d] \,\middle|\, x_{\text{MIMO}}[c] = a_j, \mathcal{H}\right). \qquad (8.26)
\end{aligned}$$

The algorithm for solving (8.26) is given in Algorithm 1 in Section 6.4. The overall complexity of MP detection for the MIMO case is of the order $O(NMQSn_{\text{R}})$, which is much lower than that of the LMMSE detector in Section 8.3.1.

8.3.3 Maximum-ratio combining detector

The LMMSE and MP detection methods are still of high complexity. To reduce the detection complexity, we extend the MRC detection from the SISO case in Section 6.5.1 to the MIMO case, where ZP-OTFS is considered. Similar to the study in Section 6.5.1, the MRC detection for MIMO ZP-OTFS can be easily extended to the other OTFS variants.

For ease of derivation, as adopted for the SISO case in Chapter 6, we consider N ZPs, each of length $L_{\text{ZP}} = l_{\max}$, to be part of an extended OTFS frame of size $M'N$, where $M' = M + l_{\max}$. At the receiver, the ZPs are not discarded, but utilized for the MRC detection. Also, we assume $L = |\mathcal{L}^{(r,t)}|$, for all r, t.

8.3.3.1 Delay-Doppler domain MRC detection

With a ZP in each block, we set the transmit and received symbol vectors as

$$\mathbf{x}^{(t)} = [(\mathbf{x}_0^{(t)})^T, \ldots, (\mathbf{x}_{M'-1}^{(t)})^T\,]^T,$$
$$\mathbf{y}^{(r)} = [(\mathbf{y}_0^{(r)})^T, \ldots, (\mathbf{y}_{M'-1}^{(r)})^T\,]^T \tag{8.27}$$

with subvectors $\mathbf{x}_m^{(t)}, \mathbf{y}_m^{(r)} \in \mathbb{C}^{N\times 1}$ for $m = 0, \ldots, M' - 1$, and the zero padding

$$\mathbf{x}_m^{(t)} = \mathbf{0} \quad \text{for} \quad m = M, \ldots, M' - 1.$$

Note that the symbol vectors in (8.27) include the ZPs, which differ from those in (8.17) which excluded the ZPs. With the effect of the ZP, the delay-Doppler input–output relation in (8.18) can be modified as

$$\mathbf{y}_{m+l}^{(r)} = \mathbf{K}_{m+l,l}^{(r,t)} \cdot \mathbf{x}_m^{(t)} + \mathbf{v}_{m,l}^{(r,t)} + \mathbf{z}_{m+l}^{(r)} \tag{8.28}$$

for $m = 0, \ldots, M - 1$ and $l \in \mathcal{L}^{(r,t)}$, where $\mathbf{K}_{m+l,l}^{(r,t)}$ are the submatrices of the extended delay-Doppler channel matrix $\mathbf{H}^{(r,t)} \in \mathbb{C}^{NM'\times NM'}$, $\mathbf{w}_{m+l}^{(r)}$ is the AWGN vector, and

$$\mathbf{v}_{m,l}^{(r,t)} = \underbrace{\sum_{l'\in\mathcal{L}^{(r,t)}, l'\neq l} \mathbf{K}_{m+l,l'}^{(r,t)} \cdot \mathbf{x}_{m+l-l'}^{(t)}}_{\text{interdelay interference}} + \underbrace{\sum_{t'\neq t} \sum_{l'\in\mathcal{L}^{(r,t')}} \mathbf{K}_{m+l,l'}^{(r,t')} \cdot \mathbf{x}_{m+l-l'}^{(t')}}_{\text{interantenna interference}} \tag{8.29}$$

are the $N \times 1$ interference vectors containing both interdelay and interantenna interference in the delay-Doppler grid, where $\mathbf{x}_{m+l-l'}^{(t)} = 0$ for $m + l - l' < 0$.

Here, we aim at estimating $\mathbf{x}_m^{(t)}$ only for $m = 0, \ldots, M - 1$, since $\mathbf{x}_m^{(t)} = \mathbf{0}$ for $m = M, \ldots, M' - 1$ are the ZP. We start with estimating the interference $\mathbf{v}_{m,l}^{(r,t)}$ in each branch, which can be done iteratively by substituting the latest estimates of symbol vectors in (8.29).

To simplify notation, we *omit* the iteration index in the following steps. Let $\hat{\mathbf{v}}_{m,l}^{(r,t)}$ be the estimated interference term in (8.29) using the latest symbol vector estimates. Removing $\hat{\mathbf{v}}_{m,l}^{(r,t)}$ from the received symbol vector $\mathbf{y}_{m+l}^{(r)}$

yields

$$\mathbf{b}_{m,l}^{(r,t)} = \mathbf{y}_{m+l}^{(r)} - \hat{\mathbf{v}}_{m,l}^{(r,t)}, \tag{8.30}$$

where $m = 0, \ldots, M-1$ and $l \in \mathcal{L}^{(r,t)}$. Substituting (8.28) in (8.30) yields

$$\mathbf{b}_{m,l}^{(r,t)} = \mathbf{K}_{m+l,l}{}^{(r,t)}\mathbf{x}_m^{(t)} + \underbrace{\mathbf{v}_{m,l}^{(r,t)} - \hat{\mathbf{v}}_{m,l}^{(r,t)}}_{\Delta\hat{\mathbf{v}}_{m,l}^{(r,t)}} + \mathbf{z}_{m+l}^{(r)}, \tag{8.31}$$

where $\Delta\hat{\mathbf{v}}_{m,l}^{(r,t)}$ denotes the residual interference due to estimation errors on $\mathbf{x}_m^{(t)}$ and can be reduced by improving symbol vector estimates iteratively. To achieve this goal, we perform MRC of the estimates $\mathbf{b}_{m,l}^{(r,t)}$ to improve the SINR at the output of the combiner, followed by symbol-by-symbol QAM demapping to make the hard decision on the QAM symbols. In summary, the MRC estimate of the m-th symbol vector of the t-th transmit antenna in each iteration is given by

$$\mathbf{c}_m^{(t)} = (\mathbf{D}_m^{(t)})^{-1} \cdot \sum_{r=1}^{n_\mathrm{R}} \sum_{l \in \mathcal{L}} \mathbf{K}_{m+l,l}^{(r,t)\dagger} \cdot \mathbf{b}_{m,l}^{(r,t)}, \quad t = 1, \ldots n_\mathrm{T}, \tag{8.32}$$

where

$$\mathbf{D}_m^{(t)} = \sum_{r=1}^{n_\mathrm{R}} \sum_{l \in \mathcal{L}} \mathbf{K}_{m+l,l}^{(r,t)\dagger} \cdot \mathbf{K}_{m+l,l}^{(r,t)} \tag{8.33}$$

and the hard decision maximum likelihood (ML) estimates at the end of each iteration are given by

$$\hat{\mathbf{x}}_m^{(t)}[n] \leftarrow \mathcal{D}(\mathbf{c}_m^{(t)}[n]) \triangleq \arg\min_{a_j \in \mathbb{A}} \left| a_j - \mathbf{c}_m^{(t)}[n] \right| \tag{8.34}$$

for $m = 0, \ldots, M-1$, $n = 0, \ldots, N-1$, and $t = 1, \ldots, n_\mathrm{T}$, where $\mathbb{A} = \{a_j, j = 1, \ldots, Q\}$ is the QAM alphabet set and $\mathcal{D}(\cdot)$ is the hard decision operation using the ML criterion.

8.3.3.2 Reduced complexity delay-time domain implementation

The complexity of MRC detection in the delay-Doppler domain can be reduced by removing redundant operations and taking advantage of the special structure of the OTFS channel matrix. Since multiplying an $N \times N$ matrix by a vector costs N^2 complex multiplications (CMs), computing $\hat{\mathbf{v}}_{m,l}^{(r,t)}$ using (8.29) incurs $(n_\mathrm{T}L - 1)N^2$ CMs.

To avoid direct computation of $\hat{\mathbf{v}}_{m,l}^{(r,t)}$, we shall conduct the following steps. To make them clear, we now *include the iteration index* in the following equations.

Using the previously estimated symbol vector $\hat{\mathbf{x}}_m^{(t)\{i-1\}}$, adding and subtracting $\mathbf{K}_{m+l,l}^{(r,t)} \cdot \hat{\mathbf{x}}_m^{(t)\{i-1\}}$ in (8.30) yields

$$\mathbf{b}_{m,l}^{(r,t)\{i\}} = \mathbf{K}_{m+l,l}^{(r,t)} \cdot \hat{\mathbf{x}}_m^{(t)\{i-1\}} + \underbrace{\mathbf{y}_{m+l}^{(r)} - \sum_{r=1}^{n_{\mathrm{R}}} \sum_{l \in \mathcal{L}^{(r,t)}} \mathbf{K}_{m+l,l}^{(r,t)} \cdot \hat{\mathbf{x}}_m^{(t)\{i-1\}}}_{\Delta \mathbf{y}_{m+l}^{(r)}}, \tag{8.35}$$

where $\Delta \mathbf{y}_{m+l}^{(r)} \in \mathbb{C}^{N \times 1}$ is the residual error plus noise (REPN) due to estimation errors on $\mathbf{x}_m^{(t)}$. Substituting (8.35) in (8.32), we obtain the MRC estimate of the symbol vector in (8.32) as

$$\mathbf{c}_m^{(t)\{i\}} = \hat{\mathbf{x}}_m^{(t)\{i-1\}} + (\mathbf{D}_m^{(t)})^{-1} \cdot \sum_{r=1}^{n_{\mathrm{R}}} \sum_{l \in \mathcal{L}^{(r,t)}} (\mathbf{K}_{m+l,l}^{(r,t)})^{\dagger} \cdot \Delta \mathbf{y}_{m+l}^{(r)}. \tag{8.36}$$

For the first iteration, we can assume all $\hat{\mathbf{x}}_m^{(t)\{0\}} = \mathbf{0}_N$, which implies that the reconstruction error term $\Delta \mathbf{y}_m^{(r)} = \mathbf{y}_m^{(r)}$. In the i-th iteration, we only need to update the $n_{\mathrm{R}} L$ reconstruction error vectors $\Delta \mathbf{y}_{m+l}^{(r)}$ for $l \in \mathcal{L}(r,t)$ and $r = 1, \ldots, n_{\mathrm{R}}$ as

$$\Delta \mathbf{y}_{m+l}^{(r)} \leftarrow \Delta \mathbf{y}_{m+l}^{(r)} - \mathbf{K}_{m+l,l}^{(r,t)} \cdot \Delta \mathbf{x}_m^{(t)\{i\}}, \tag{8.37}$$

where

$$\Delta \mathbf{x}_m^{(t)\{i\}} = \mathbf{c}_m^{(t)\{i\}} - \hat{\mathbf{x}}_m^{(t)\{i-1\}}. \tag{8.38}$$

Note that in (8.38) we can replace $\mathbf{c}_m^{(t)\{i\}}$ by the hard estimate $\hat{\mathbf{x}}_m^{(t)\{i\}}$. After updating $\Delta \mathbf{y}_{m+l}^{(r)}$, we repeat the same procedure for remaining m and t.

The core steps of the MRC detector in (8.32) and (8.33) are hence reduced to (8.36), (8.37), and (8.38). The steps in (8.36) and (8.37) involve $n_{\mathrm{T}}(n_{\mathrm{R}} L + 1)MN^2$ and $n_{\mathrm{T}} n_{\mathrm{R}} LMN^2$ CMs per iteration, respectively. This complexity, even though lower than that of the MMSE detection and comparable to that of the MP detection, is still high. However, thanks to the special structure of the OTFS channel matrix, the complexity can be further reduced by taking advantage of the fact that $\mathbf{K}_{m,l}^{(r,t)}$ are circulant matrices, which can be diagonalized in the delay-time domain.

From the delay-time input–output relation in (8.19) and (8.20), the operations in (8.37) and (8.38) can be performed with much lower complexity in this domain as

$$\Delta \tilde{\mathbf{x}}_m^{(t)\{i\}}[n] = \frac{1}{\tilde{\mathbf{d}}_m^{(t)}[n]} \sum_{r=1}^{n_{\mathrm{R}}} \sum_{l \in \mathcal{L}^{(r,t)}} \tilde{\boldsymbol{\nu}}_{m+l,l}^{(r,t)*}[n] \Delta \tilde{\mathbf{y}}_{m+l}^{(r)}[n] \tag{8.39}$$

and the reconstruction error at all branches $l \in \mathcal{L}^{(r,t)}$ for $r = 1, \ldots, n_R$ can be updated as

$$\Delta\tilde{\mathbf{y}}_{m+l}^{(r)}[n] \leftarrow \Delta\tilde{\mathbf{y}}_{m+l}^{(r)}[n] - \tilde{\boldsymbol{\nu}}_{m+l,l}^{(r,t)}[n]\Delta\tilde{\mathbf{x}}_m^{(t)\{i\}}[n] \tag{8.40}$$

for $n = 0, \ldots, N-1$, where

$$\tilde{\mathbf{d}}_m^{(t)}[n] = \sum_{r=1}^{n_R} \sum_{l \in \mathcal{L}^{(r,t)}} |\tilde{\boldsymbol{\nu}}_{m+l,l}^{(r,t)}[n]|^2. \tag{8.41}$$

The estimated information symbol at the end of the i-th iteration (in the delay-Doppler domain) is given by

$$\hat{\mathbf{x}}_m^{(t)\{i\}} = \hat{\mathbf{x}}_m^{(t)\{i-1\}} + \mathbf{F}_N \cdot \Delta\tilde{\mathbf{x}}_m^{(t)\{i\}}. \tag{8.42}$$

To improve convergence speed, instead of using the delay-Doppler information symbol estimate, we can use a weighted average of such estimates with hard decision as

$$\mathbf{c}_m^{(t)\{i\}} \leftarrow (1-\delta)\hat{\mathbf{x}}_m^{(t)\{i\}} + \delta\mathcal{D}(\hat{\mathbf{x}}_m^{(t)\{i\}}). \tag{8.43}$$

The value of δ can be slightly increased every iteration as the norm of the residual error decreases and the hard decision estimates become more reliable. The iterations are stopped when the residual error does not decrease any more or when the maximum number of iterations is reached. All the steps in the reduced complexity delay-time MIMO MRC detection are summarized in Algorithm 1.

8.3.3.3 MRC detection complexity

We now discuss the complexity of Algorithm 1. The operations in steps 10 and 13 require $(2n_RL + 1)$ CMs per iteration per information symbol. At the end of each iteration the two N-FFTs in steps 19 and 21 of Algorithm 1 require $2N \log_2 N$ CMs per symbol vector per iteration. Computing $\tilde{\mathbf{d}}_m^{(t)}[n]$ in (8.41) requires n_RL CMs per information symbol. Computing the reconstruction error in step 25 requires n_RL CMs per symbol vector per iteration. Assuming S iterations are required, the overall number of CMs required for detecting all information symbols using Algorithm 1 is $n_TNM[n_RL + S(3n_RL + 2\log_2 N + 1)]$. This is significantly lower than the complexity of MP detection ($O(n_Tn_RNMSP^2Q)$) and LMMSE detection ($O((n_Tn_RNM)^3)$).

8.4 MIMO-OTFS channel estimation

In this section, we extend the embedded pilot channel estimation for SISO-OTFS in Chapter 7 to MIMO-OTFS with n_T transmit and n_R receive

Algorithm 1: MRC MIMO detection algorithm.

Received signal input: $\tilde{\mathbf{y}}_m^{(r)}$ for $m = 0, \ldots, M' - 1$ and all r

Channel input: $\tilde{\boldsymbol{\nu}}_{m,l}^{(r,t)}$ for $m = 0, \ldots, M' - 1$, $l \in \mathcal{L}^{(r,t)}$ and all t, r

$\tilde{\mathbf{d}}_m^{(t)}$ for $m = 0, \ldots, M - 1$, $l \in \mathcal{L}^{(r,t)}$ and all t, r

Initialize: $\hat{\mathbf{x}}_m^{(t)\{0\}} = \mathbf{0}_N$ for all t and $m = 0, \ldots, M - 1$

$\Delta\tilde{\mathbf{y}}_m^{(r)} = \tilde{\mathbf{y}}_m^{(r)}$ for all t, r and $m = 0, \ldots, M' - 1$

for *i=1:max iterations* **do**
    for $m = 0 : M - 1$ **do**
        for $n = 0 : N - 1$ **do**
            for $t = 1 : n_T$ **do**
                $\Delta\tilde{\mathbf{x}}_m^{(t)\{i\}}[n] = (\tilde{\mathbf{d}}_m^{(t)}[n])^{-1} \times \sum_{r=1}^{n_R} \sum_{l \in \mathcal{L}^{(r,t)}} \tilde{\boldsymbol{\nu}}_{m+l,l}^{(r,t)*}[n] \Delta\tilde{\mathbf{y}}_{m+l}^{(r)}[n]$
                for $r = 1 : n_R$ **do**
                    for $l \in \mathcal{L}^{(r,t)}$ **do**
                        $\Delta\tilde{\mathbf{y}}_{m+l}^{(r)}[n] \leftarrow \Delta\tilde{\mathbf{y}}_{m+l}^{(r)}[n] - \tilde{\boldsymbol{\nu}}_{m+l,l}^{(r,t)}[n] \Delta\mathbf{x}_m^{(t)\{i\}}[n]$
                    end
                end
            end
        end
    end
    for $m = 0 : M - 1$ **do**
        for $t = 1 : n_T$ **do**
            $\hat{\mathbf{x}}_m^{(t)\{i\}} = \hat{\mathbf{x}}_m^{(t)\{i-1\}} + \mathbf{F}_N \cdot \Delta\tilde{\mathbf{x}}_m^{(t)\{i\}}$
            $\hat{\mathbf{c}}_m^{(t)\{i\}} \leftarrow (1 - \delta)\hat{\mathbf{x}}_m^{(t)\{i\}} + \delta\mathcal{D}(\hat{\mathbf{x}}_m^{(t)\{i\}})$
            $\tilde{\mathbf{c}}_m^{(t)\{i\}} = \mathbf{F}_N^{\dagger} \cdot \hat{\mathbf{c}}_m^{(t)\{i\}}$
            $\Delta\tilde{\mathbf{x}}_m^{(t)\{i\}} = \tilde{\mathbf{c}}_m^{(t)\{i\}} - \mathbf{F}_N^{\dagger} \cdot \hat{\mathbf{x}}_m^{(t)\{i\}}$
            for $r = 1 : n_R$ **do**
                for $l \in \mathcal{L}^{(r,t)}$ **do**
                    $\Delta\tilde{\mathbf{y}}_{m+l}^{(r)} \leftarrow \Delta\tilde{\mathbf{y}}_{m+l}^{(r)} - \tilde{\boldsymbol{\nu}}_{m+l,l}^{(r,t)} \circ \tilde{\Delta}\mathbf{x}_m^{(t)\{i\}}$
                end
            end
        end
    end
end

Output: $\mathcal{D}(\hat{\mathbf{x}}_m^{(t)\{i\}})$ for all m, t.

antennas. For ease of explanation, assuming integer delay and Doppler taps, we recall $l_{\max}$ (the maximum channel delay tap among all channels) and $k_{\max} = \max(\max_l(\mathcal{K}_l))$, where $\mathcal{K}_l$ is the set of integer Doppler taps for the l-th delay path in all channels.

In the point-to-point SISO-OTFS case, the embedded pilot channel estimation scheme allocates a single pilot symbol, zero guard symbols, and data symbols in the delay-Doppler grid. The pilot is surrounded by the guard symbols that occupy the rectangular region of size $(2l_{\max} + 1) \times (4k_{\max} + 1)$.

We can extend this idea to estimate the channel between each antenna pair in MIMO-OTFS. Hence, we embed the pilot symbol for each transmit antenna in the delay-Doppler grid, with sufficient spacing in-between, so that they do not interfere with each other at the receiver. At each receive antenna, the channels are simultaneously estimated from the nonoverlapping received pilots within a frame.

Fig. 8.2 illustrates two embedded pilot channel estimation schemes for 2×2 MIMO-OTFS, where pilot, data, and guard symbols are arranged in the delay-Doppler grid. In the figure, the red (dark gray in print version) box shows the pilot and guard symbol region for each transmit antenna and the extra guard symbols outside the red (dark gray in print version) box are reserved to avoid interference from pilots of other transmit antennas.

Specifically, in Fig. 8.2(a), the pilot placement is shown for moderate Doppler spread channels (i.e., $2k_{\max} + 1 < N/n_{\mathrm{T}}$). The pilots can be placed on the same delay index but are separated by $2k_{\max}$ along the Doppler dimension to guarantee no interference at the receiver.

For the high Doppler spread cases (i.e., $2k_{\max} + 1 = N$), as shown in Fig. 8.2(b), full guard symbols that occupy the entire Doppler dimension can be adopted, while the pilots can be placed on the same Doppler index but are separated by $2l_{\max}$ along the delay dimension to avoid interference at the receiver.

In summary, MIMO-OTFS channel estimation in Fig. 8.2(a) and (b) requires the pilot and guard symbol overhead

$$\frac{(2l_{\max} + 1)(2k_{\max} + 1)n_{\mathrm{T}}}{MN} \quad \text{and} \quad \frac{((n_{\mathrm{T}} + 1)l_{\max} + n_{\mathrm{T}}) \times N}{MN},$$

respectively. Other arrangements are possible with the reduced size of pilot regions. For example, we can transmit partially overlapping pilot regions, but this will result in pilot collision. Separation of transmitter pilots at the receiver depending on the channel characteristics needs to be investigated further.

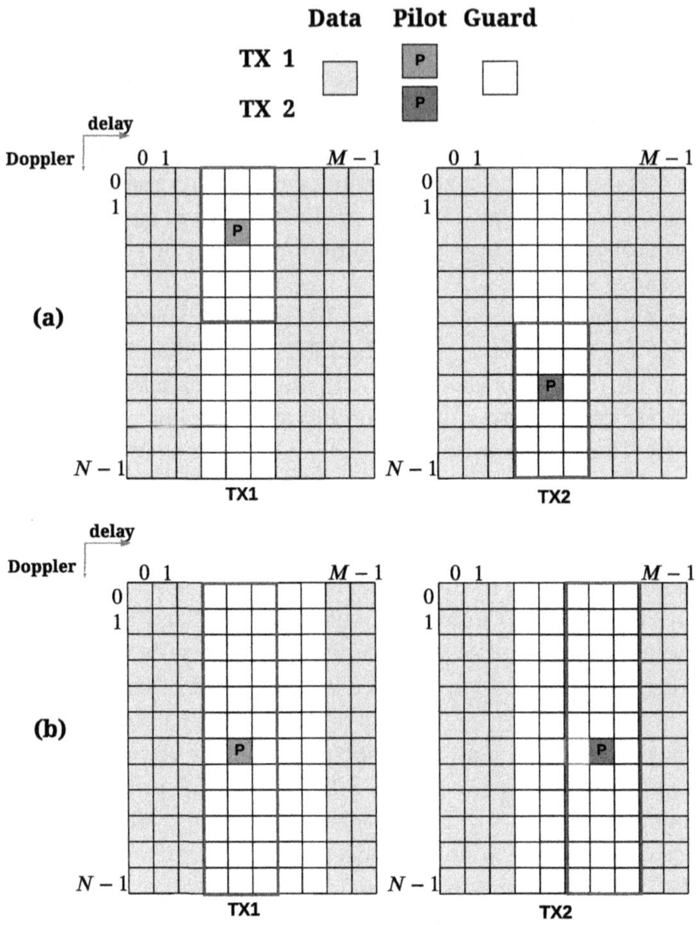

FIGURE 8.2 MIMO-OTFS embedded pilot channel estimation schemes for (a) moderate and (b) high Doppler spread channel estimation.

8.5 Multiuser OTFS channel estimation

In this section, we extend the above embedded pilot channel estimation for MIMO-OTFS to multiuser OTFS, where a single antenna is adopted at each user and the receive terminal, respectively.

In multiuser OTFS uplink, we allocate resource blocks (RBs) of individual users in different positions in the delay-Doppler grid for channel estimation and data transmission, with sufficient spacing between adjacent pilots to avoid interference at the receiver. Each user's RB contains a single pilot symbol, zero guard symbols, and data symbols. Fig. 8.3 illustrates two embedded pilot channel estimation schemes for a four-user

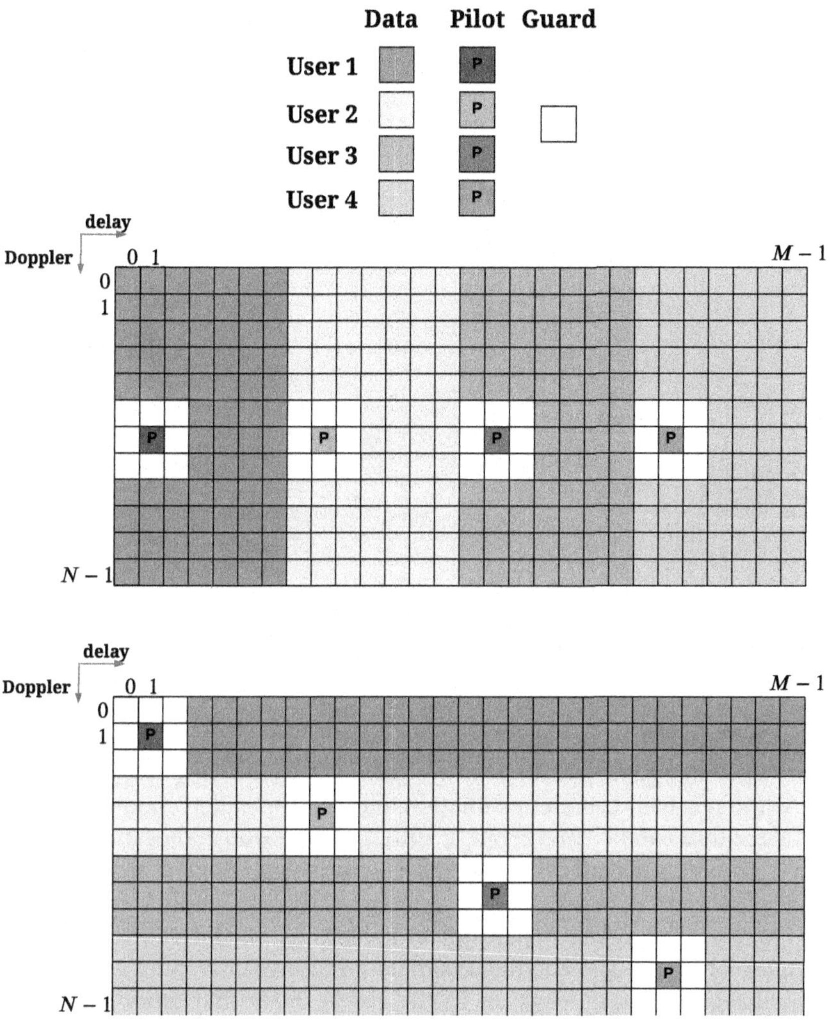

FIGURE 8.3 Multiuser OTFS embedded pilot channel estimation schemes for uplink communication with four users.

uplink case. The RBs of individual users occupy four nonoverlapping regions in the delay-Doppler grid for data transmissions, where each region is marked with its distinct color.

In the first scheme, these four nonoverlapping regions split along the delay dimension in the delay-Doppler grid. The pilots can be placed on the same Doppler index but with at least $2l_{\max}$ guard symbols along the delay axis to guarantee there is no interference at the receiver.

To save pilot overhead, we assume $2l_{\max}$ guard symbols are placed along the delay axis between adjacent pilots. In this case, each pilot and

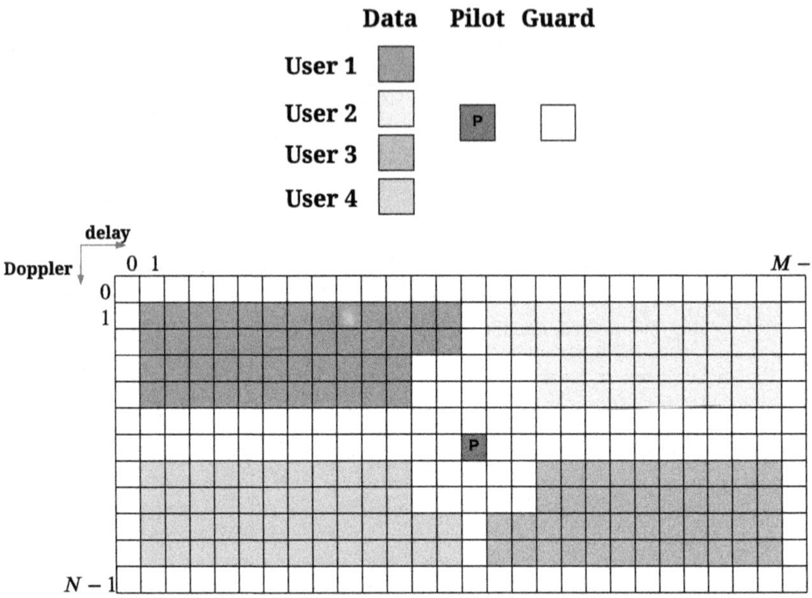

FIGURE 8.4 Multiuser OTFS embedded pilot channel estimation scheme for downlink communication with four users.

guard symbol region has size $(2l_{\max}+1)(4k_{\max}+1)$, and the total number of the pilot and guard symbols for N_u users is $((N_u+1)l_{\max}+N_u)(4k_{\max}+1)$. At the receiver, channel estimation and multiuser detection using the MP algorithm can be conducted. If orthogonal resource allocation is preferred, then guard symbols can be inserted between adjacent data regions.

In the second scheme, these four nonoverlapping regions split along the Doppler dimension, where pilots are placed on different delay and Doppler locations. The channel estimation requires each pilot and guard symbol region of size $(2l_{\max}+1)(4k_{\max}+1)$. Note that other pilot and guard symbol arrangements can be made with reduced overhead at the cost of higher transmit complexity.

In multiuser OTFS downlink, we consider a single antenna base-station, transmitting a single pilot symbol, zero guard symbols, and data symbols to users, similar to the SISO-OTFS case. The pilot signal is used by all the downlink users to estimate the broadcast channel. The remaining delay-Doppler grid locations are allocated for guard symbols and data symbols. Fig. 8.4 shows an embedded pilot channel estimation scheme for a four-user downlink case, when orthogonal resource allocation is required. A single pilot is surrounded by guard symbols to guarantee there is no interference between pilot and data symbols, leading to the pilot region of size $(2l_{\max}+1)(4k_{\max}+1)$. Moreover, extra guard symbols are placed between users' data symbols to avoid interference at the receiver. In summary, the arrangement of pilot, guard, and data symbols can be designed according to the requirements of resource allocation schemes, overhead, and transmit/receive complexity.

8.6 Numerical results and discussion

In this section, we present simulation results on the bit error rate (BER) of ZP-OTFS for the SISO and MIMO cases and compare them with the CP-OFDM performance. We adopt the following system parameters: a sub-carrier spacing of 15 kHz, a carrier frequency of 4 GHz, and 4, 16, 64-QAM signaling. We consider a five-path channel model with a uniform power delay profile with the set of integer delay taps $\mathcal{L}^{(r,t)} = \{0, 1, 2, 3, 4\}$ for all possible (r, t) antenna pairs and the single Doppler shift for each path generated according to the Jakes spectrum as $\nu_{\max} \cos\theta_i$, where $\nu_{\max}$ is the maximum Doppler shift determined by the UE speed and θ_i is uniformly distributed over $[-\pi, \pi]$. For each SNR point of the BER performance, we simulate 10^4 OTFS frames. For channel estimation, the pilot power at each transmitter is taken as $E_p = N(2l_{\max}+1)E_s$, where E_s is the average energy of the independently generated QAM symbols.

The code for generating the channel for each antenna pair is given in MATLAB® codes 6 and 7 in Appendix C.

Fig. 8.5 compares the BER performance of OTFS and OFDM using 4-QAM signaling for the SISO, 2 × 2 MIMO, and 3 × 3 MIMO cases. Each scheme adopts a block-wise LMMSE detector for a frame size of $N = M = 32$ at a UE speed of 120 km/h. The BERs of both OFDM and OTFS improve with the number of transmit and receive antennas, but OTFS offers significant improvement over OFDM in high mobility MIMO channels.

Fig. 8.6 compares the BER performance for SISO-OTFS, 2 × 2 MIMO-OTFS, and 3 × 3 MIMO-OTFS with 4-QAM signaling using LMMSE, MP, and MRC detectors, respectively, with maximum 30 iterations. Each scheme adopts a frame size of $N = M = 32$ at a UE speed of 120 km/h. We observe that the low complexity MRC detector offers better performance than LMMSE and MP detectors at moderate to high SNRs.

Fig. 8.7 shows the BER performance of SISO-OTFS, 2 × 2 MIMO-OTFS, and 3 × 3 MIMO-OTFS with the MRC detector for 16, 64-QAM symbols with a frame size $M = N = 32$ and a UE speed of 120 km/h. The excellent error performance of the MRC detector is highlighted by the fact that it converges for 16-QAM with $\delta = 0.125$ and maximum 25 iterations and for 64-QAM with $\delta = 0.05$ and maximum 40 iterations at linear complexity.

Figs. 8.8 and 8.9 illustrate the BER performance of 2 × 2 and 4 × 4 MIMO-OTFS systems at the SNR of 15 dB, using the embedded pilot channel estimation scheme in Fig. 8.2(a) with pilot spacing of N/n_T, with LMMSE

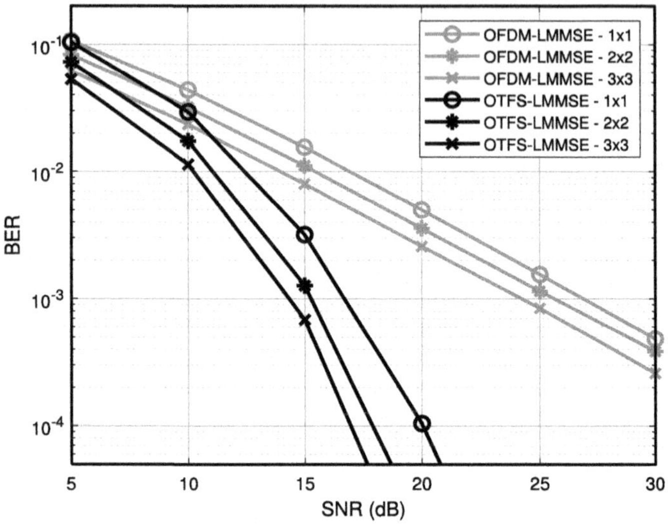

FIGURE 8.5 OTFS vs. OFDM BER performance using 4-QAM for block-wise LMMSE detection for a frame size of $N = M = 32$ for different number of antennas.

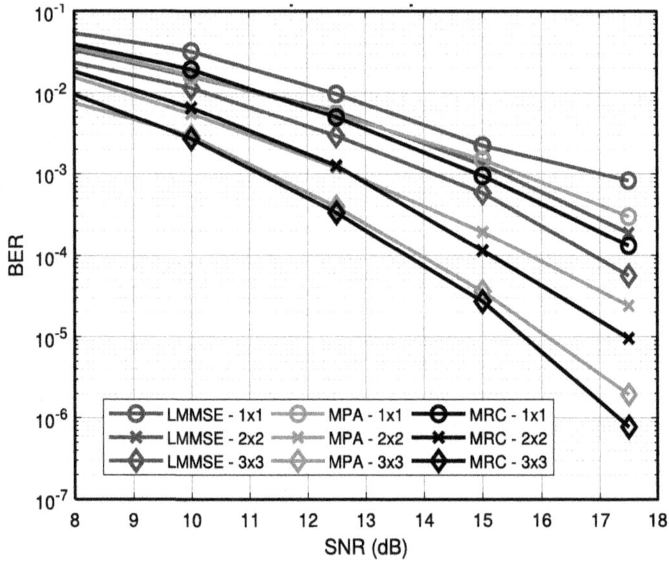

FIGURE 8.6 OTFS uncoded BER performance using 4-QAM for block-wise LMMSE vs. MP vs. MRC detection for a frame size of $N = M = 32$ for different number of antennas.

and MRC detections at the receiver. In Fig. 8.8, the ideal cases with perfect CSI using the LMMSE and MRC detections are also plotted in dashed

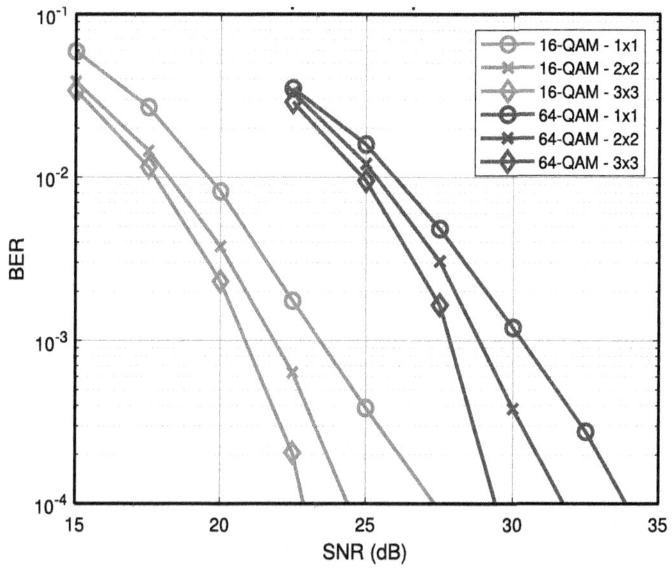

FIGURE 8.7 OTFS BER performance with MRC detector for a frame size of $N = M = 32$ for different modulation sizes and number of antennas.

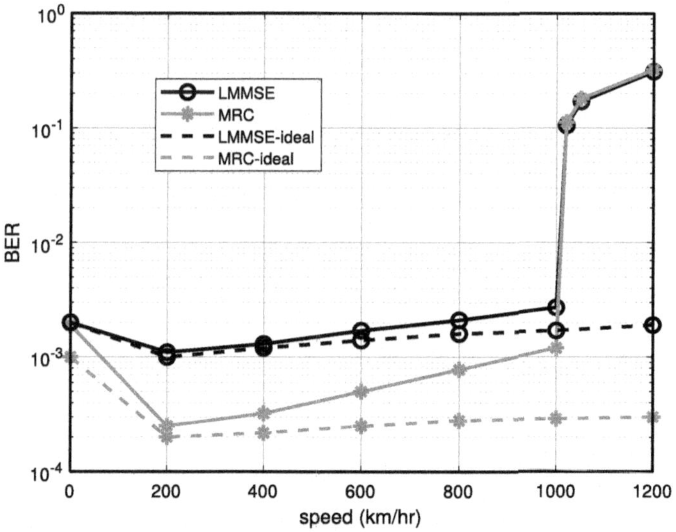

FIGURE 8.8 2×2 MIMO-OTFS BER performance using 4-QAM for block-wise LMMSE vs. MRC detector for a frame size of $N = M = 32$ for different UE speeds with channel estimation in Fig. 8.2(a) at an SNR of 15 dB.

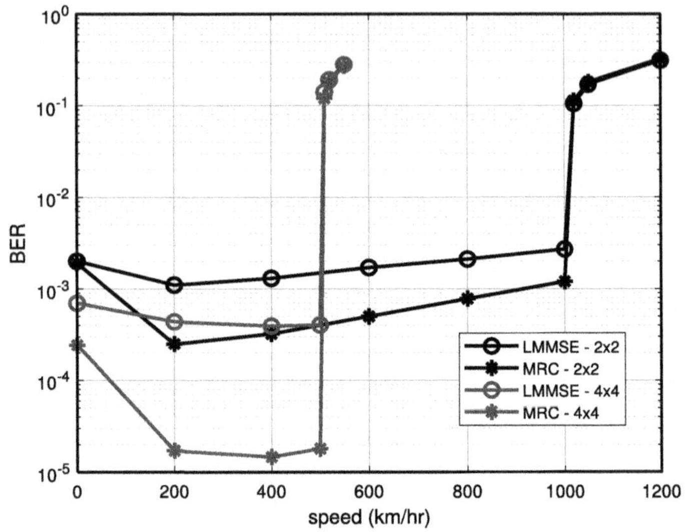

FIGURE 8.9 2×2 vs. 4×4 MIMO-OTFS BER performance using 4-QAM for block-wise LMMSE vs. MRC (maximum 30 iterations) detection for a frame size of $N = M = 32$ for different UE speeds with channel estimation in Fig. 8.2(a) at an SNR of 15 dB.

lines. In both figures with perfect and estimated CSIs, the BERs using the iterative MRC (maximum 30 iterations) are better than those using the LMMSE detector.

In both figures, we observe that the channel estimation scheme in Fig. 8.2(a) is effective for a maximum UE speed of 1000 km/h and 550 km/h for 2×2 and 4×4 MIMO-OTFS systems. It becomes ineffective when the UE speed goes above 1000 km/h for 2×2 MIMO and 550 km/h for 4×4 MIMO, due to insufficient pilot spacing $N/n_T < 2k_{\max} + 1$. As discussed in Section 8.4, for such high Doppler cases, the solution is to simply adopt the scheme in Fig. 8.2(b), where we place pilots in different delay locations and full guard symbols along the entire Doppler dimension to accommodate high Doppler spreads.

8.7 Bibliographical notes

OTFS modulation was proposed by Hadani *et al.* in the 2017 IEEE Wireless Communications and Networking Conference [1]. The input–output relations and analysis of MIMO-OTFS systems can be found in [2,3]. The MP detection presented in Section 8.3.2 was proposed for MIMO-OTFS in [2]. The MIMO-OTFS detection and/or channel estimation methods presented in the chapter were proposed in [2,4–6]. Other MIMO and multiuser OTFS channel estimation and/or detection were discussed in [5,7–16]. Other aspects such as coding, diversity, beamforming, etc., for MIMO and/or multiuser OTFS can be found in [17–22]. Recent research on the MRC detection for MIMO-OTFS and correlated MIMO-OTFS can be found in [23].

References

[1] R. Hadani, S. Rakib, M. Tsatsanis, A. Monk, A.J. Goldsmith, A.F. Molisch, R. Calderbank, Orthogonal time frequency space modulation, in: 2017 IEEE Wireless Communications and Networking Conference (WCNC), 2017, pp. 1–6.

[2] M. Kollengode Ramachandran, A. Chockalingam, MIMO-OTFS in high-Doppler fading channels: Signal detection and channel estimation, in: 2018 IEEE Global Communications Conference (GLOBECOM), 2018, pp. 206–212.

[3] A. RezazadehReyhani, A. Farhang, M. Ji, R.R. Chen, B. Farhang-Boroujeny, Analysis of discrete-time MIMO OFDM-based orthogonal time frequency space modulation, in: 2018 IEEE International Conference on Communications (ICC), 2018, pp. 1–6.

[4] G.D. Surabhi, A. Chockalingam, Low-complexity linear equalization for 2×2 MIMO-OTFS signals, in: 2020 IEEE 21st International Workshop on Signal Processing Advances in Wireless Communications (SPAWC), 2020, pp. 1–5.

[5] P. Raviteja, K. Tran, Y. Hong, Embedded pilot-aided channel estimation for OTFS in delay-Doppler channels, IEEE Transactions on Vehicular Technology 68 (5) (2019) 4906–4917, https://doi.org/10.1109/TVT.2019.2906357.

[6] P. Raviteja, K. Phan, Y. Hong, E. Viterbo, Embedded delay-Doppler channel estimation for orthogonal time frequency space modulation, in: 2018 IEEE 88th Vehicular Technology Conference (VTC-Fall), 2018, pp. 1–6.

[7] W. Shen, L. Dai, S. Han, I.C. Lin, R.W. Heath, Channel estimation for orthogonal time frequency space (OTFS) massive MIMO, in: 2019 IEEE International Conference on Communications (ICC), 2019, pp. 1–6.

[8] W. Shen, L. Dai, J. An, P. Fan, R.W. Heath, Channel estimation for orthogonal time frequency space (OTFS) massive MIMO, IEEE Transactions on Signal Processing 67 (16) (2019) 4204–4217, https://doi.org/10.1109/TSP.2019.2919411.

[9] Y. Liu, S. Zhang, F. Gao, J. Ma, X. Wang, Uplink-aided high mobility downlink channel estimation over massive MIMO-OTFS system, IEEE Journal on Selected Areas in Communications 38 (9) (2020) 1994–2009, https://doi.org/10.1109/JSAC.2020.3000884.

[10] O.K. Rasheed, G.D. Surabhi, A. Chockalingam, Sparse delay-Doppler channel estimation in rapidly time-varying channels for multiuser OTFS on the uplink, in: 2020 IEEE 91st Vehicular Technology Conference (VTC2020-Spring), 2020, pp. 1–5.

[11] Y. Liu, S. Zhang, F. Gao, J. Ma, X. Wang, Uplink-aided high mobility downlink channel estimation over massive MIMO-OTFS system, IEEE Journal on Selected Areas in Communications 38 (9) (2020) 1994–2009, https://doi.org/10.1109/JSAC.2020.3000884.

[12] B.C. Pandey, S.K. Mohammed, P. Raviteja, Y. Hong, E. Viterbo, Low complexity precoding and detection in multi-user massive MIMO OTFS downlink, IEEE Transactions on Vehicular Technology 70 (5) (2021) 4389–4405, https://doi.org/10.1109/TVT.2021.3061694.

[13] D. Shi, W. Wang, L. You, X. Song, Y. Hong, X. Gao, G. Fettweis, Deterministic pilot design and channel estimation for downlink massive MIMO-OTFS systems in presence of the fractional Doppler, IEEE Transactions on Wireless Communications 20 (11) (2021) 7151–7165, https://doi.org/10.1109/TWC.2021.3081164, in press.

[14] M. Li, S. Zhang, F. Gao, P. Fan, O.A. Dobre, A new path division multiple access for the massive MIMO-OTFS networks, IEEE Journal on Selected Areas in Communications 39 (4) (2021) 903–918, https://doi.org/10.1109/JSAC.2020.3018826.

[15] A. Naikoti, A. Chockalingam, Low-complexity delay-Doppler symbol DNN for OTFS signal detection, in: 2021 IEEE 93rd Vehicular Technology Conference (VTC2021-Spring), 2021, pp. 1–6.
[16] P. Singh, H.B. Mishra, R. Budhiraja, Low-complexity linear MIMO-OTFS receivers, in: 2021 IEEE International Conference on Communications Workshops (ICC Workshops), 2021, pp. 1–6.
[17] R.M. Augustine, G.D. Surabhi, A. Chockalingam, Space-time coded OTFS modulation in high-Doppler channels, in: 2019 IEEE 89th Vehicular Technology Conference (VTC2019-Spring), 2019, pp. 1–6.
[18] V.S. Bhat, S.G. Dayanand, A. Chockalingam, Performance analysis of OTFS modulation with receive antenna selection, IEEE Transactions on Vehicular Technology 70 (4) (2021) 3382–3395, https://doi.org/10.1109/TVT.2021.3063546.
[19] Z. Ding, Robust beamforming design for OTFS-NOMA, IEEE Open Journal of the Communications Society 1 (2019) 33–40, https://doi.org/10.1109/OJCOMS.2019.2953574.
[20] V. Khammammetti, S.K. Mohammed, OTFS-based multiple-access in high Doppler and delay spread wireless channels, IEEE Wireless Communications Letters 8 (2) (2019) 528–531, https://doi.org/10.1109/LWC.2018.2878740.
[21] A.K. Sinha, S.K. Mohammed, P. Raviteja, Y. Hong, E. Viterbo, OTFS based random access preamble transmission for high mobility scenarios, IEEE Transactions on Vehicular Technology 69 (12) (2020) 15078–15094, https://doi.org/10.1109/TVT.2020.3034130.
[22] G.D. Surabhi, R.M. Augustine, A. Chockalingam, On the diversity of uncoded OTFS modulation in doubly-dispersive channels, IEEE Transactions on Wireless Communications 18 (6) (2019) 3049–3063, https://doi.org/10.1109/TWC.2019.2909205.
[23] T. Thaj, E. Viterbo, Low-complexity linear diversity-combining detector for MIMO-OTFS, IEEE Wireless Communications Letters (2021) 1, https://doi.org/10.1109/LWC.2021.3125986.

CHAPTER

9

Conclusions and future directions

OUTLINE

Chapter points

- OTFS key advantages.
- Other research directions of interest.

The larger the island of knowledge, the longer the shoreline of wonder.
Ralph W. Sockman

This book presents a family of delay-Doppler communication systems and reveals how they could efficiently operate on high mobility wireless channels. We started with the modeling and the delay-Doppler representation of a wireless high mobility channel. Then we focused on the delay-

Delay-Doppler Communications
https://doi.org/10.1016/B978-0-32-385028-5.00017-7

Doppler modulation OTFS, which transmits information symbols in the delay-Doppler domain to combat severe Doppler shifts in high mobility scenarios. We analyzed the interaction of OTFS signals with high mobility channels and extended our analysis to the variants of OTFS.

To understand the fundamental idea underpinning OTFS, we presented its relation to the well-known Zak transform. Capitalizing on the fact that OTFS modulation/demodulation is equivalent to the discrete Zak transform, we showed how to simplify the analysis using the Zak transform properties.

In later chapters, we introduced various detection and channel estimation techniques operating in different domains. We also presented a real-time software-defined radio (SDR) implementation highlighting the impacts and mitigation strategies for hardware impairments. Finally, we extended our analysis of detection and channel estimation schemes from single input single output (SISO) to multiple input multiple output (MIMO) and multiuser communications.

This chapter summarizes the key advantages of delay-Doppler modulation and lists the pros and cons of the OTFS variants in terms of transmit power, detection complexity, and normalized spectral efficiency (NSE), including the pilot overhead required for channel estimation. The comparison is helpful for readers to choose the suitable OTFS variant for some specific design requirements. Finally, we list several research directions of interest for future exploration.

9.1 OTFS key advantages

4G wireless communications have adopted the OFDM waveform as the air interface for slowly time-varying (quasistatic) multipath channels. OFDM uses multiple subcarriers to carry information, and all subcarriers remain orthogonal at the receiver by using a cyclic prefix (CP) to combat the intersymbol interference (ISI), introduced by the multipath channel delay spread. This enables the use of a low complexity single tap equalizer for detection. Since the channel is slowly varying, it can be estimated by dispersing pilots in multiple OFDM symbols, and the channel estimates for all subcarriers can be obtained through simple time-frequency domain interpolation.

In high mobility scenarios, wireless channels are doubly selective, where multipath effects result in ISI and Doppler shifts. Thus, the channel changes within an OFDM symbol, which incurs a loss of orthogonality among subcarriers. This, in turn, causes severe intercarrier interference (ICI) and performance degradation. The low complexity detection and channel estimation for OFDM are very challenging in such scenarios. Multiple Dopplers are hard to equalize, subchannel gains are not equal, and the worst one determines the performance. We may need to use powerful channel coding to compensate for such drawbacks. Further, to estimate such rapidly varying channels, OFDM would require a very high pilot overhead, impacting the spectral efficiency.

Different from OFDM, OTFS multiplexes information symbols over two-dimensional orthogonal basis functions in the delay-Doppler domain rather than the time-frequency domain. Such two-dimensional orthogonal basis functions are specifically designed to combat the dynamics of the time-varying multipath channels. As a result, OTFS converts the fading, time-varying multipath channel into a sparse and slowly time-varying channel [1]. This delay-Doppler representation captures the geometry of the wireless environments, as discussed in Chapter 2.

When observed in the time domain, the orthogonal basis functions of OTFS span an NM-dimensional space. At the receiver, these functions are all affected by the same scaling of the channel but interfere with other basis functions for a high mobility multipath channel with P paths. Consequently, the orthogonality at the receiver is lost, and some ISI from other symbols appears. Fortunately, the small number of paths $P \ll NM$ can still allow us to equalize this interference easily. For example, as shown in Chapter 6, we can use very low complexity maximum ratio combining (MRC) detection to achieve performance comparable to that of message passing (MP) detection.

In high mobility channels, the traditional time-frequency domain does not capture the separate propagation paths affected by the Doppler shifts. In contrast, the delay-Doppler domain enables a sparse and slowly time-varying channel representation, resulting in a simple estimation of the parameters of each propagation path. A single pilot symbol can be used to estimate the delay, Doppler shifts, and gain of each path for the entire OTFS frame. Assuming there are P propagation paths with distinct integer delays and Doppler shifts, a pilot symbol in isolation transmitted in the delay-Doppler domain produces P distinct nonzero samples at the receiver, which define the delay-Doppler channel response. For fractional Doppler shifts, it is possible to estimate the delay-Doppler channel response through a larger number of distinct nonzero samples at the receiver.

Motivated by this, the embedded pilot delay-Doppler channel estimation is presented in Chapter 7, where a pilot, guard, and data symbols are specifically arranged in an OTFS frame to avoid interference between pilot and data at the receiver. Alternatively, the delay-time domain channel estimation can be used, which has the advantage of requiring fewer parameters to represent the channel owing to the sparsity offered by the delay-time domain. Then an interpolation is performed to reconstruct the time domain channel coefficients for the entire frame for each delay (see details in Chapter 7).

Overall, OTFS exhibits excellent performance not only in high mobility channels but also in static multipath channels. It was shown in [2] that, for static multipath channels, the system structure of OTFS is equivalent to the vector OFDM (V-OFDM) in [3] (also known as the asymmetric OFDM (A-OFDM) in [4,5]). With a low complexity MP or MRC detection, OTFS achieves a significant performance gain compared with the above schemes with ZF and MMSE detection. The superior performance of OTFS as compared to widely used OFDM makes it a potential waveform candidate for 6G wireless communications.

9.2 Pros and cons of OTFS variants

In Chapter 4, we introduced four OTFS variants and compared their key design constraints in Table 4.4, including transmit power (P_{T}), channel sparsity, and NSE. It was shown that ZP/RZP-OTFS have the least transmit power.

In Chapter 6, we presented the MRC detection and its complexity for the OTFS variants. We observed that ZP-OTFS has the least detection complexity thanks to some zero channel submatrices ($\mathbf{K}_{m,l} = \mathbf{0}_N$ for $m < l$), while RCP/RZP-OTFS have the highest complexity due to noncirculant submatrices $\mathbf{K}_{m,l}$.

In Chapter 7, we studied the effect of channel estimation on the NSE. It was shown in Table 7.1 that, for high mobility channels with full guard channel estimation, ZP-OTFS has the highest NSE.

To help readers choose the suitable OTFS variant for a specific design requirement, we summarize in Table 9.1 their pros and cons in terms of transmit power (P_{T}), the MRC detection complexity, and NSE[1] including pilot overhead for channel estimation in low and high mobility environments. When comparing the NSEs for low and high mobility channels, we adopt the embedded pilot channel estimation in Chapter 7 with reduced and full guard symbols, respectively.

From Table 9.1, it can be concluded that ZP-OTFS offers the best solution in high mobility channels due to low transmit power, low complexity of detection, and high NSE. Additionally, ZP-OTFS enables the use of low complexity delay-time domain channel estimation, as discussed in Chapter 7.

9.3 Other research directions

In this section, we present a number of research directions of interest to the research community working on OTFS. We would like to point out

[1]Here, NSE is the ratio between the original spectral efficiency and $\log_2 Q$ for Q-QAM signaling.

TABLE 9.1 OTFS variants pros and cons (P_T from Table 4.4, MRC detection from Section 6.5.6, NSE from Table 7.1).

	Transmit power (P_T)	**MRC detection complexity**	**NSE reduced guard (low mobility)**	**NSE full guard (high mobility)**
CP	moderate	low	low	low
ZP	lowest	lowest	moderate	high
RCP	low	moderate	high	moderate
RZP	lowest	moderate	high	moderate

that the Zak transform is a powerful tool that may be helpful in analyzing and solving some of the delay-Doppler communication problems.

9.3.1 Channel estimation and PAPR reduction

The embedded pilot channel estimation in Chapter 7 does not estimate the fractional Doppler shifts in each delay path. Instead, it estimates the full Doppler channel response samples. Alternatively, we can estimate the individually fractional Doppler shifts to fit the full Doppler channel response. However, this nonlinear estimation process can lead to an inaccurate channel estimate under certain channel conditions such as closely spaced fractional Doppler shifts. Additionally, noise can significantly affect the accuracy of the estimate. Hence, estimating fractional Doppler shifts is an interesting open problem to be studied.

For the embedded pilot channel estimation, another issue is the peak-to-average power ratio (PAPR), which determines the efficiency of the transmitter power amplifier (PA). A low PAPR indicates that the PA operates efficiently in the linear region, while a high PAPR causes the PA to operate in the nonlinear region, leading to degraded performance. It was shown in [6] that the maximum PAPR of OTFS grows linearly with the number of time slots N, rather than the number of subcarriers M like in OFDM. Thus, OTFS has better PAPR than OFDM for $N < M$. However, when using the embedded pilot delay-time channel estimation, a single pilot is placed in the delay-Doppler domain, surrounded by guard symbols. Such placement can lead to a reduced complexity channel estimation over traditional methods in [7], but it incurs high PAPR due to spike impulses in the time domain, produced by the pilot placed in the delay-Doppler domain.

One solution is to reduce pilot power so that the pilot spikes fall well below the peaks of the time domain samples. Another solution is to place a pilot sequence (with the same total power as a single pilot) in the delay-Doppler domain [8,9]. However, channel estimation with such a pilot sequence has higher complexity. Nevertheless, given its advantages in regard to PAPR, it can be considered an interesting problem to be explored.

9.3.2 Channels with fast time-varying delay-Doppler paths

Future wireless networks may involve communications between vehicles moving at very high speeds. For very long frames, this may cause the geometry of the physical channel to change within the duration of a frame. One such scenario is that when a high speed vehicle accelerates, decelerates, or turns a corner within a frame duration (tens of milliseconds), as for a racing car communicating with the pit crew. The delay, Doppler, and propagation gain then become functions of time. Omitting the noise term, the received signal can be written in terms of the time-varying delay-Doppler coefficients (gain $g_i(t)$, delay spread $\tau_i(t)$, Doppler shift $\nu_i(t)$ per path) as

$$r(t) = \sum_{i=1}^{P} g_i(t) e^{j2\pi \nu_i(t)(t-\tau_i(t))} s(t-\tau_i(t)). \tag{9.1}$$

Such a model may also be used for underwater acoustic communications with moving terminals [10,11], where the multipath is due to signal reflections from the surface and the bottom of the sea, and the time-varying multipath parameters are caused by the relative motion between transmitter and receiver and sea wave propagation [12].

The Zak transform techniques presented in Chapter 5 may be explored to simplify the problem of analyzing the fast time-varying delay-Doppler channel. In addition, for such channels, the time variability makes it difficult to accurately estimate the delay-Doppler channel parameters. Specifically, if we adopt OTFS with the embedded pilot channel estimation in Chapters 7 and 8, this estimation technique may not be able to capture the time-varying nature of the channel coefficients. Accurate modeling of how the delay-Doppler parameters in (9.1) change with respect to time in a frame duration is a challenging problem to be explored.

9.3.3 Multiuser communications

Consider the OTFS multiuser uplink, where multiple users communicate with the base-station (BS) simultaneously, and each user places data in a distinct portion of the delay-Doppler domain. For detection methods, if sufficient guard symbols are placed between the portions of the delay-Doppler domain, then single user detection can be used [13]. Otherwise, the joint multiuser detection can be conducted via MP [13]. Additionally, the MRC detection can be applied for a joint multiuser detection, similar to the multiple input single output (MISO) case discussed in Chapter 8. An alternative allocation of the delay-Doppler domain resources can be designed to limit the bandwidth or time slots assigned to the users [14].

If channel knowledge is available on the users' side, precoding or beamforming can be applied to make users' signals orthogonal and reduce detection complexity [15,16]. Hybrid beamforming, which includes a combination of precoding and analog beamforming by adjusting the phases of the transmit antennas, can offer a good solution to increase user capacity in shared spaces. Alternatively, intelligent reflecting surface (IRS) [17–22] can also be used to shape the geometry of the multipath channel. Therefore, the use of OTFS in conjunction with precoding, beamforming, and IRS for the multiuser uplink are exciting topics to be studied.

Another challenge in multiuser communication is the uplink channel estimation [13,23], as the number of pilots that can be transmitted in the given bandwidth and time duration is limited by the frame size and the delay and Doppler spread of the channel. More pilots may be required to accommodate more users, and some mechanism to avoid pilot collision or decouple overlapping pilots must be developed. This, along with efficient multiuser detection in very high mobility channels, remains an open problem for OTFS.

Nonorthogonal multiple access (NOMA) is known to support a large number of users to utilize time and frequency resources efficiently [24]. Since OTFS can be interpreted as a two-dimensional CDMA, where information symbols are spread over the entire time and frequency plane [1,25], it is natural to combine OTFS and NOMA for multiuser communications in high mobility scenarios. Very recently, OTFS-NOMA has been proposed in [24,26–29]. These works have focused on the development of OTFS-NOMA schemes, beamforming techniques, and detection methods (see details in [24,26–29]). It was demonstrated that OTFS-NOMA achieves improved spectral efficiency and better error performance for users with different mobility profiles. Future studies could explore the applications of NOMA to the OTFS variants and investigate channel estimation and detection techniques.

9.3.4 Massive MIMO-OTFS

Massive MIMO is one of the key technologies for 5G wireless communications. Massive MIMO adopts a large number of antennas to offer huge improvements in throughput and energy efficiency [30–33]. For a high mobility multiuser downlink, where the BS communicates with a large number of users in motion, OTFS can be used in conjunction with massive MIMO to improve the error performance [8,9,15,34–36].

The performance of massive MIMO-OTFS for multiuser downlink can be analyzed in the delay-Doppler-space domain (or delay-Doppler-angle domain) [8,9,34,36]. However, a major challenge lies in the channel estimation when operating in the frequency division duplex (FDD) mode [8,9]. It is known that in time division duplex (TDD) mode [36], channel knowledge can be obtained by uplink training thanks to uplink–downlink channel reciprocity. In contrast, in FDD mode [8,9,34], downlink channel estimation is more challenging. Embedded pilot- and pilot sequence-aided channel estimation were proposed in [8,9] by exploiting three-dimensional channel sparsity, but with a relatively high pilot overhead for limited channel conditions. Low overhead downlink channel estimation of massive MIMO-OTFS is an open problem to be explored.

9.3.5 OTFS for RadCom

RadCom refers to joint radar and communications via a common waveform. RadCom has found a wide range of applications, both in modern civilian and commercial areas. For example, in emerging intelligent transportation applications, RadCom systems offer both communication links to other vehicles and active environment sensing functionalities, enabling cooperative interactions between all vehicles on the road [37]. Other applications of RadCom systems are unsurprisingly found in aeronautical and military areas. However, despite much research, there remain limitations of the current RadCom (and radar) systems to support emerging applications such as intelligent automotive systems with high mobility and dense traffic environments requiring very high data rates, ultrareliability, and ultralow latency communications. Under these conditions, it is challenging to develop suitable waveforms to simultaneously satisfy the requirements of radar sensing and data communications.

As a promising waveform for high mobility environments, OTFS has been initially studied in [38,39] for RadCom, including the input–output relation, detection, and radar estimation. It was shown that OTFS-based radar processing not only exhibits the inherent advantages of multicarrier modulation but also provides additional benefits of improved radar capability, such as longer range, faster tracking rate, etc. These results inspire the potential applications of OTFS in RadCom. A possible research direction is the exploration of RadCom with the different OTFS variants introduced in Chapter 4.

9.3.6 Orthogonal time sequency multiplexing and precoding design

OTFS offers excellent performance in both static and high mobility multipath channels. The performance gain is a result of spreading the information symbols via ISFFT on two-dimensional orthogonal basis functions that span the entire time and bandwidth resource, thereby exploiting time-frequency diversity. We can think of ISFFT in OTFS as a two-dimensional precoding of the time-frequency domain.

In [16,40], it was shown that any two-dimensional unitary transformation in the time-frequency domain can offer the same error performance as the ISFFT for OTFS. This opens up many opportunities to choose time-frequency precoding without sacrificing performance.

One example is the Walsh–Hadamard transform (WHT) waveform proposed in [41] (similar to CP-OTFS by replacing FFT with WHT), where a time domain linear minimum mean squared error with parallel interference cancelation (LMMSE-PIC) is used at the receiver. The other example is the recently proposed orthogonal time sequency multiplexing (OTSM) [42,43], which spreads the information symbols in time and frequency, using FFT along frequency and WHT ($\mathbf{W}_N \in \mathbb{Z}^{N\times N}$) along time. A low complexity delay-time (or delay-Doppler) MRC detection and time domain channel estimation are proposed.

In general, the unitary precoding in the time-frequency domain can be written as $\mathbf{X}_{\text{tf}} = \mathbf{V}_{\text{f}} \cdot \mathbf{X} \cdot \mathbf{V}_{\text{t}}$, where $\mathbf{X}_{\text{tf}} \in \mathbb{C}^{M\times N}$ are the time-frequency samples matrix generated from the information symbols matrix $\mathbf{X} \in \mathbb{C}^{M\times N}$, and the unitary matrices $\mathbf{V}_{\text{f}} \in \mathbb{C}^{M\times M}$ and $\mathbf{V}_{\text{t}} \in \mathbb{C}^{N\times N}$ define different waveforms. For example, OTFS uses the ISFFT with $\mathbf{V}_{\text{f}} = \mathbf{F}_M$ and $\mathbf{V}_{\text{t}} = \mathbf{F}_N^\dagger$, while OTSM uses $\mathbf{V}_{\text{f}} = \mathbf{F}_M$ and $\mathbf{V}_{\text{t}} = \mathbf{W}_N$. Both schemes use $\mathbf{V}_{\text{f}} = \mathbf{F}_M$ which cancels out with the IFFT of the Heisenberg transform as shown below,

$$\mathbf{s} = \text{vec}(\mathbf{F}_M^\dagger \cdot \mathbf{X}_{\text{tf}}) = \text{vec}(\mathbf{F}_M^\dagger \cdot \mathbf{F}_M \cdot \mathbf{X} \cdot \mathbf{V}_{\text{t}}) = \text{vec}(\mathbf{X} \cdot \mathbf{V}_{\text{t}}), \tag{9.2}$$

generating the time domain vector $\mathbf{s}$ for transmission. The choice of $\mathbf{V}_{\text{f}} = \mathbf{F}_M$ converts the time-frequency two-dimensional unitary precoded waveform into a one-dimensional unitary precoded single carrier waveform. Furthermore, the choice of $\mathbf{V}_{\text{t}}$ affects the computational complexity of the precoding operation. For example, the use of WHT ($\mathbf{V}_{\text{t}} = \mathbf{W}_N$) in OTSM requires only additions and subtractions, while the use of $\mathbf{V}_{\text{t}} = \mathbf{F}_N^\dagger$ in OTFS requires complex multiplications with higher complexity.

On the other hand, the choice of the precoding matrix $\mathbf{V}_{\text{t}}$ affects the channel sparsity, which determines detection complexity. Assuming channels with integer delay and Doppler taps only, OTFS using $\mathbf{V}_{\text{t}} = \mathbf{F}_N^\dagger$ provides the sparsest representation of the channel. However, when a channel has fractional Doppler shifts, such a representation may not be as sparse. Instead, the use of WHT in OTSM leads to a similar sparsity (number of delay-Doppler taps) with reduced computational complexity. Another alternative is to use the discrete cosine transform (DCT), which may offer more sparsity (reduced number of delay-Doppler taps) than IDFT and WHT. The DCT is known to concentrate most of the energy in a few coefficients, but the sparse representation of the channel offered by DCT is unexplored in the literature. A broad research direction is to devise unitary precoding guaranteeing sparse channel representation for various systems over different types of channels.

9.3.7 Machine learning for OTFS

Machine learning technology has found its way into various sectors and industries, aiming at optimizing system performance using massive training datasets. Deep neural network (DNN)-based machine learning is a promising tool for solving many physical layer communication problems, including channel estimation and prediction and signal detection with imperfect or no knowledge of channel state information (see [44–50] and references therein).

Very recently, machine learning techniques have been applied to OTFS, particularly for channel estimation and detection in [51,52]. In [51], a DNN-based transceiver architecture was proposed to learn the delay-Doppler channel and detect the information symbols using a single DNN trained for this purpose. In [52], two-dimensional convolutional neural network (CNN)-based signal detection was proposed for OTFS. A data augmentation technique based on the MP algorithm was used to improve the learning ability of the proposed method.

In conclusion, practical design, implementation, and validations of machine learning techniques for OTFS are still in their infancy and are a key direction for future research.

References

[1] R. Hadani, S. Rakib, M. Tsatsanis, A. Monk, A.J. Goldsmith, A.F. Molisch, R. Calderbank, Orthogonal time frequency space modulation, in: 2017 IEEE Wireless Communications and Networking Conference (WCNC), 2017, pp. 1–6.

[2] P. Raviteja, E. Viterbo, Y. Hong, OTFS performance on static multipath channels, IEEE Wireless Communications Letters 8 (3) (2019) 745–748, https://doi.org/10.1109/LWC.2018.2890643.

[3] X.G. Xia, Precoded and vector OFDM robust to channel spectral nulls and with reduced cyclic prefix length in single transmit antenna systems, IEEE Transactions on Communications 49 (8) (2001) 1363–1374, https://doi.org/10.1109/26.939855.

[4] J. Zhang, A.D.S. Jayalath, Y. Chen, Asymmetric OFDM systems based on layered FFT structure, IEEE Signal Processing Letters 14 (11) (2007) 812–815, https://doi.org/10.1109/LSP.2007.903230.

[5] L. Luo, J. Zhang, Z. Shi, BER analysis for asymmetric OFDM systems, in: Proc. 2008 IEEE Global Telecommunications Conference (GLOBECOM), 2008, pp. 1–6.

[6] G.D. Surabhi, R.M. Augustine, A. Chockalingam, Peak-to-average power ratio of OTFS modulation, IEEE Communications Letters 23 (6) (2019) 999–1002, https://doi.org/10.1109/LCOMM.2019.2914042.

[7] M. Ramachandran, A. Chockalingam, MIMO-OTFS in high-Doppler fading channels: signal detection and channel estimation, in: 2018 IEEE Global Communications Conference (GLOBECOM), 2018, pp. 1–6.

[8] W. Shen, L. Dai, J. An, P.Z. Fan, R.W. Heath, Channel estimation for orthogonal time frequency space (OTFS) massive MIMO, IEEE Transactions on Signal Processing 67 (16) (2019) 4204–4217, https://doi.org/10.1109/TSP.2019.2919411.

[9] D. Shi, W. Wang, L. You, X. Song, Y. Hong, X. Gao, G. Fettweis, Deterministic pilot design and channel estimation for downlink massive MIMO-OTFS systems in presence of the fractional Doppler, IEEE Transactions on Wireless Communications 20 (11) (2021) 7151–7165, https://doi.org/10.1109/TWC.2021.3081164.

[10] T. Ebihara, G. Leus, Doppler-resilient orthogonal signal-division multiplexing for underwater acoustic communication, IEEE Journal of Oceanic Engineering 41 (2) (2016) 408–427, https://doi.org/10.1109/JOE.2015.2454411.

[11] T. Ebihara, H. Ogasawara, G. Leus, Underwater acoustic communication using multiple-input–multiple-output Doppler-resilient orthogonal signal division multiplexing, IEEE Journal of Oceanic Engineering 45 (4) (2020) 1594–1610, https://doi.org/10.1109/JOE.2019.2922094.
[12] S.H. Byun, S.M. Kim, Y.K. Lim, W. Seong, Time-varying underwater acoustic channel modeling for moving platform, in: Oceans 2017, 2007, pp. 1–6.
[13] P. Raviteja, K.T. Phan, Y. Hong, Embedded pilot-aided channel estimation for OTFS in delay-Doppler channels, IEEE Transactions on Vehicular Technology 68 (5) (2019) 4906–4917, https://doi.org/10.1109/TVT.2019.2906357.
[14] V. Khammammetti, S.K. Mohammed, OTFS-based multiple-access in high Doppler and delay spread wireless channels, IEEE Wireless Communications Letters 8 (2) (2019) 528–531, https://doi.org/10.1109/LWC.2018.2878740.
[15] B.C. Pandey, S.K. Mohammed, P. Raviteja, Y. Hong, E. Viterbo, Low complexity precoding and detection in multi-user massive MIMO OTFS downlink, IEEE Transactions on Vehicular Technology 70 (5) (2021) 4389–4405, https://doi.org/10.1109/TVT.2021.3061694.
[16] T. Zemen, D. Loschenbrand, M. Hofer, C. Pacher, B. Rainer, Orthogonally precoded massive MIMO for high mobility scenarios, IEEE Access 7 (2019) 132979–132990, https://doi.org/10.1109/ACCESS.2019.2941316.
[17] S. Hu, F. Rusek, O. Edfors, The potential of using large antenna arrays on intelligent surfaces, in: 2017 IEEE 85th Vehicular Technology Conference (VTC Spring), 2017, pp. 1–6.
[18] S. Hu, F. Rusek, O. Edfors, Beyond massive MIMO: the potential of data transmission with large intelligent surfaces, IEEE Transactions on Signal Processing 66 (10) (2019) 2746–2758, https://doi.org/10.1109/TSP.2018.2816577.
[19] Q. Wu, R. Zhang, Intelligent reflecting surface enhanced wireless network: joint active and passive beamforming design, in: 2018 IEEE Global Communications Conference (GLOBECOM), 2018, pp. 1–6.
[20] Q. Wu, R. Zhang, Intelligent reflecting surface enhanced wireless network via joint active and passive beamforming, IEEE Transactions on Wireless Communications 18 (11) (2019) 5394–5409, https://doi.org/10.1109/TWC.2019.2936025.
[21] E. Basar, M.D. Renzo, J.D. Rosny, M. Debbah, M. Alouini, R. Zhang, Wireless communications through reconfigurable intelligent surfaces, IEEE Access 7 (2019) 116753–116773, https://doi.org/10.1109/ACCESS.2019.2935192.
[22] Q. Wu, R. Zhang, Towards smart and reconfigurable environment: intelligent reflecting surface aided wireless network, IEEE Communications Magazine 58 (1) (2020) 106–112, https://doi.org/10.1109/MCOM.001.1900107.
[23] O.K. Rasheed, G.D. Surabhi, A. Chockalingam, Sparse delay-Doppler channel estimation in rapidly time-varying channels for multiuser OTFS on the uplink, in: 91st IEEE Vehicular Technology Conference, VTC Spring 2020, Antwerp, Belgium, May 25–28, 2020, IEEE, 2020, pp. 1–5.
[24] Z. Ding, R. Schober, P. Fan, H.V. Poor, OTFS-NOMA: An efficient approach for exploiting heterogenous user mobility profiles, IEEE Transactions on Communications 67 (11) (2019) 7950–7965, https://doi.org/10.1109/TCOMM.2019.2932934.
[25] P. Raviteja, K.T. Phan, Y. Hong, E. Viterbo, Interference cancellation and iterative detection for orthogonal time frequency space modulation, IEEE Transactions on Wireless Communications 17 (10) (2018) 6501–6515, https://doi.org/10.1109/TWC.2018.2860011.
[26] Z. Ding, Robust beamforming design for OTFS-NOMA, IEEE Open Journal of the Communications Society 1 (2019) 33–40, https://doi.org/10.1109/OJCOMS.2019.2953574.
[27] Y. Ge, Q. Deng, P.C. Ching, Z. Ding, OTFS signaling for uplink NOMA of heterogeneous mobility users, IEEE Transactions on Communications 69 (5) (2019) 3147–3161, https://doi.org/10.1109/TCOMM.2021.3059456.

[28] K. Deka, A. Thomas, S. Sharma, OTFS-SCMA: a code-domain NOMA approach for orthogonal time frequency space modulation, IEEE Transactions on Communications 69 (8) (2021) 5043–5058, https://doi.org/10.1109/TCOMM.2021.3075237.
[29] A. Chatterjee, V. Rangamgari, S. Tiwari, S.S. Das, Nonorthogonal multiple access with orthogonal time-frequency space signal transmission, IEEE Systems Journal 15 (1) (2021) 383–394, https://doi.org/10.1109/JSYST.2020.2999470.
[30] L. Marzetta, Noncooperative cellular wireless with unlimited num Bers of base station antennas, IEEE Transactions on Wireless Communications 9 (11) (2010) 3590–3600, https://doi.org/10.1109/TWC.2010.092810.091092.
[31] F. Rusek, D. Persson, B.K. Lau, E.G. Larsson, T.L. Marzetta, O. Edfors, F. Tufvesson, Scaling up MIMO: opportunities and challenges with very large arrays, IEEE Signal Processing Magazine 30 (1) (2010) 40–60, https://doi.org/10.1109/MSP.2011.2178495.
[32] H.Q. Ngo, E.G. Larsson, T.L. Marzetta, Energy and spectral efficiency of very large multiuser MIMO systems, IEEE Transactions on Communications 61 (4) (2013) 1436–1449, https://doi.org/10.1109/MSP.2011.2178495.
[33] E.G. Larsson, O. Edfors, F. Tufvesson, T.L. Marzetta, Massive MIMO for next generation wireless systems, IEEE Communications Magazine 52 (2) (2014) 186–195, https://doi.org/10.1109/MCOM.2014.6736761.
[34] Y. Liu, S. Zhang, F. Gao, J. Ma, X. Wang, Uplink-aided high mobility downlink channel estimation over massive MIMO-OTFS system, IEEE Journal on Selected Areas in Communications 38 (9) (2020) 1994–2009, https://doi.org/10.1109/JSAC.2020.3000884.
[35] Y. Shan, F. Wang, Low-complexity and low-overhead receiver for OTFS via large-scale antenna array, IEEE Transactions on Vehicular Technology 70 (6) (2021) 5703–5718, https://doi.org/10.1109/TVT.2021.3072667.
[36] M. Li, S. Zhang, F. Gao, P. Fan, O.A. Dobre, Low-complexity and low-overhead receiver for OTFS via large-scale antenna array, IEEE Journal on Selected Areas in Communications 39 (4) (2021) 903–918, https://doi.org/10.1109/JSAC.2020.3018826.
[37] C. Sturm, W. Wiesbeck, Waveform design and signal processing aspects for fusion of wireless communications and radar sensing, Proceedings of the IEEE 99 (7) (2011) 1236–1259, https://doi.org/10.1109/JPROC.2011.2131110.
[38] P. Raviteja, K.T. Phan, Y. Hong, E. Viterbo, Orthogonal time frequency space (OTFS) modulation based radar system, in: 2019 IEEE Radar Conference (RadarConf), 2021, pp. 1–6.
[39] L. Gaudio, M. Kobayashi, G. Caire, G. Colavolpe, On the effectiveness of OTFS for joint radar parameter estimation and communication, IEEE Transactions on Wireless Communications 19 (9) (2020) 5951–5965, https://doi.org/10.1109/TWC.2020.2998583.
[40] T. Zemen, M. Hofer, D. Loschenbrand, C. Pacher, Iterative detection for orthogonal precoding in doubly selective channels, in: 2018 IEEE 29th Annual International Symposium on Personal, Indoor and Mobile Radio Communications (PIMRC), 2018, pp. 1–6.
[41] R. Bomfin, A. Nimr, M. Chafii, G. Fettweis, A robust and low-complexity Walsh–Hadamard modulation for doubly-dispersive channels, IEEE Communications Letters 25 (3) (2021) 897–901, https://doi.org/10.1109/LCOMM.2020.3034429.
[42] T. Thaj, E. Viterbo, Orthogonal time sequency multiplexing modulation, in: 2021 IEEE Wireless Communications and Networking Conference (WCNC), 2021.
[43] T. Thaj, E. Viterbo, Y. Hong, Orthogonal time sequency multiplexing modulation: analysis and low-complexity receiver design, IEEE Transactions on Wireless Communications 20 (12) (2021) 7842–7855, https://doi.org/10.1109/TWC.2021.3088479.
[44] T. O'Shea, J. Hoydis, An introduction to deep learning for the physical layer, IEEE Transactions on Cognitive Communications and Networking 3 (4) (2017) 563–575, https://doi.org/10.1109/TCCN.2017.2758370.
[45] H. Ye, G.Y. Li, B. Juang, Power of deep learning for channel estimation and signal detection in OFDM systems, IEEE Wireless Communications Letters 7 (1) (2018) 114–117, https://doi.org/10.1109/LWC.2017.2757490.

[46] S. Dorner, S. Cammerer, J. Hoydis, S.t. Brink, Deep learning based communication over the air, IEEE Journal of Selected Topics in Signal Processing 12 (1) (2018) 132–143, https://doi.org/10.1109/JSTSP.2017.2784180.

[47] N. Farsad, A. Goldsmith, An introduction to deep learning for the physical layer, IEEE Transactions on Signal Processing 66 (21) (2017) 5663–5678, https://doi.org/10.1109/TSP.2018.2868322.

[48] S. Sharma, Y. Hong, UWB receiver via deep learning in MUI and ISI scenarios, IEEE Transactions on Vehicular Technology 69 (3) (2020) 3496–3499, https://doi.org/10.1109/TVT.2020.2972510.

[49] T.V. Luong, Y. Ko, N.A. Vien, D.H.N. Nguyen, M. Matthaiou, Deep learning-based detector for OFDM-IM, IEEE Wireless Communications Letters 8 (4) (2019) 1159–1162, https://doi.org/10.1109/LWC.2019.2909893.

[50] S. Sharma, Y. Hong, A hybrid multiple access scheme via deep learning-based detection, IEEE Systems Journal 15 (1) (2021) 981–984, https://doi.org/10.1109/JSYST.2020.2975666.

[51] A. Naikoti, A. Chockalingam, Low-complexity delay-Doppler symbol DNN for OTFS signal detection, in: 2021 IEEE 93rd Vehicular Technology Conference (VTC2021-Spring), 2021, pp. 1–6.

[52] Y.K. Enku, B. Bai, F. Wan, C.U. Guyo, I.N. Tiba, C. Zhang, S. Li, Two-dimensional convolutional neural network based signal detection for OTFS systems, IEEE Wireless Communications Letters 10 (11) (2021) 2514–2518, https://doi.org/10.1109/LWC.2021.3106039.

APPENDIX

A

Notation and acronyms

$\mathbb{R}^{M\times N}$	set of $(M \times N)$-dimensional matrices with real entries
$\mathbb{C}^{M\times N}$	set of $(M \times N)$-dimensional matrices with complex entries
$\mathbb{Z}^{M\times N}$	set of $(M \times N)$-dimensional matrices with integer entries
a	real or complex number
a^*	complex conjugation
$\mathbf{a}$	column vectors
$\mathbf{a}[n]$	n-th element of vector $\mathbf{a}$
$\mathbf{A}$	matrices
$\mathbf{A}[m,n]$	(m,n)-th entry of matrix $\mathbf{A}$
$(\cdot)^T$	transposition
$(\cdot)^\dagger$	Hermitian transposition
$\mathbf{A}^n$	n-th power of a square matrix $\mathbf{A}$
$\text{diag}(\mathbf{a})$	diagonal matrix with vector $\mathbf{a}$ on the main diagonal
$\text{circ}(\mathbf{a})$	square circulant matrix based circular shifts of column vector $\mathbf{a}$
$\text{tr}(\mathbf{A})$	trace of the square matrix $\mathbf{A}$
$\mathbf{a} = \text{vec}(\mathbf{A})$	column-wise vectorization of the $M \times N$ matrix $\mathbf{A}$ to a vector $\mathbf{a}$ of length MN
$\mathbf{A} = \text{vec}^{-1}_{M,N}(\mathbf{a})$	inverse vectorization of the vector $\mathbf{a}$ of length MN to form an $M \times N$ matrix $\mathbf{A}$
$\mathbf{I}_N$	$N \times N$ identity matrix
$\mathbf{0}_N$	column vector of N zeros
$\mathbf{1}_N$	column vector of N ones
$\circledast$	circular convolution
$\otimes$	matrix Kronecker product
$\circ$	matrix Hadamard product (element-wise multiplication)
$\oslash$	matrix Hadamard division (element-wise division)
$\mathbf{F}_N$	N-point discrete Fourier transform (DFT) matrix
$\mathbf{F}_N^\dagger$	N-point inverse discrete Fourier transform (IDFT) matrix
$\text{DFT}_N(\cdot)$	N-point discrete Fourier transform operator
$\text{IDFT}_N(\cdot)$	N-point inverse discrete Fourier transform operator
$\delta(\cdot)$	Dirac delta function

$[n]_M$	integer n modulo the integer M
$\lceil a \rceil$	ceiling function mapping a to the closest integer greater than or equal to a
$\lfloor a \rfloor$	floor function mapping a to the closest integer less than or equal to a
$\lvert \mathcal{S} \rvert$	cardinality of the set $\mathcal{S}$
$c = 3 \times 10^8$ m/s	speed of light
$z = e^{\frac{j2\pi}{MN}}$	
$\max(\cdot)$	max operation that returns the maximum value of a set of elements
$\mathrm{E}(\cdot)$	expectation operator
$\mathrm{Var}(\cdot)$	variance operator

1D	one-dimensional
2D	two-dimensional
3D	three-dimensional
AD	analog-to-digital
ADC	analog-to-digital converter
AWGN	additive white Gaussian noise
BC	broadcast channel
BER	bit error rate
CDMA	code division multiple access
CFO	carrier frequency offset
CM	complex multiplication
CP	cyclic prefix
CP-OTFS	cyclic prefix orthogonal time-frequency space
DA	digital-to-analog
DAC	digital-to-analog converter
DC	direct current
DCR	direct conversion receiver
DFE	decision feedback equalizer
DFT	discrete Fourier transform
DLL	dynamic linked libraries
DZT	discrete Zak transform
ECC	error correcting code
EGC	equal gain combining
EPA	extended pedestrian A model
ETU	extended typical urban model
EVA	extended vehicular A model
FBMC	filter bank multicarrier
FDD	frequency division duplex
FDE	frequency domain equalizer
FDMA	frequency division multiple access
FFT	fast Fourier transform
GFDM	generalized frequency division multiplexing
GSM	Global System for Mobile Communications
ICI	intercarrier interference
IDFT	inverse discrete Fourier transform

IDI	inter-Doppler interference
IDZT	inverse discrete Zak transform
IF	intermediate frequency
IFFT	inverse fast Fourier transform
IQ	in-phase and quadrature
IRS	intelligent reflecting surface
ISFT	inverse symplectic Fourier transform
ISFFT	inverse symplectic fast Fourier transform
ISI	intersymbol interference
LDPC	low density parity check
LLR	log likelihood ratio
LMMSE	linear minimum mean square error
LMMSE-PIC	linear minimum mean square error with parallel interference cancellation
LTI	linear time-invariant
MAC	multiple access channel
MAP	maximum a posteriori probability
MIMO	multiple input multiple output
MISO	multiple input single output
MMSE	minimum mean square error
MP	message passing algorithm
MRC	maximum ratio combining
NI	National Instruments
NOMA	nonorthogonal multiple access
NPI	noise plus interference
NSE	normalized spectral efficiency
OFDM	orthogonal frequency division multiplexing
OFDMA	orthogonal frequency division multiplexing access
OOB	out-of-band
OTFS	orthogonal time-frequency space
OTFS	orthogonal time-frequency shift (before 2017)
OTSM	orthogonal time sequency multiplexing
PAPR	peak-to-average power ratio
PS	parallel-to-serial
PS-OFDM	pulse shaped orthogonal frequency division multiplexing
QAM	quadrature amplitude modulation
RCP-OTFS	reduced cyclic prefix orthogonal time-frequency space
REPN	residual error plus noise
RZP-OTFS	reduced zero padding orthogonal time-frequency space
SE	spectral efficiency
SFT	symplectic Fourier transform
SFFT	symplectic fast Fourier transform
SDR	software-defined radio
SINR	signal-to-interference-plus-noise ratio
SISO	single input single output
SNR	signal-to-noise ratio

TDD	time division duplex
TDL	tapped delay line
TDMA	time division multiple access
UFMC	universal filtered multicarrier
UMTS	universal mobile telecommunications system
USRP	universal software radio peripheral
WHT	Walsh–Hadamard transform
ZP	zero padding
ZP-OTFS	zero padding orthogonal time-frequency space

APPENDIX

B

Some useful matrix properties

B.1 The DFT matrix

The N-point discrete Fourier transform (DFT) matrix is defined as

$$\mathbf{F}_N = \frac{1}{\sqrt{N}}\left\{\omega_N^{kl}\right\}_{k,l=0}^{N-1} = \frac{1}{\sqrt{N}}\begin{pmatrix} 1 & 1 & 1 & & 1 \\ 1 & \omega_N^{1\cdot 1} & \omega_N^{1\cdot 2} & \cdots & \omega_N^{(N-1)\cdot 1} \\ \vdots & & & & \vdots \\ 1 & \omega_N^{(N-1)\cdot 1} & \omega_N^{(N-1)\cdot 2} & \cdots & \omega_N^{(N-1)(N-1)} \end{pmatrix},$$

where $\omega_N = e^{-j2\pi/N}$. Given a vector $\mathbf{a}$ of length N we denote its DFT as $\tilde{\mathbf{a}} = \mathbf{F}_N\mathbf{a}$. The IDFT matrix is given by $\mathbf{F}_N^{-1} = \mathbf{F}_N^{\dagger}$ and $\mathbf{a} = \mathbf{F}_N^{\dagger}\tilde{\mathbf{a}}$.

The columns of $\mathbf{F}_N$ are harmonic discrete-time complex sinusoidal functions, which form the DFT orthonormal basis. We finally remark the symmetry property $\mathbf{F}_N^{\mathrm{T}} = \mathbf{F}_N$ and $(\mathbf{F}_N^{\dagger})^{\mathrm{T}} = \mathbf{F}_N^{\dagger}$.

B.2 Permutation matrices

A permutation is a one-to-one mapping within the set of integers $1, \ldots N$, defined as

$$\pi = \begin{pmatrix} 1 & 2 & \cdots & n \\ \pi_1 & \pi_2 & & \pi_N \end{pmatrix}.$$

We can permute the elements of a given vector by multiplying it by a *permutation matrix* $\mathbf{P}$ which has a 1 in each row i at position π_i and zero otherwise,

$$\mathbf{P}\begin{pmatrix} a_1 \\ a_2 \\ \vdots \\ a_N \end{pmatrix} = \begin{pmatrix} a_{\pi_1} \\ a_{\pi_2} \\ \vdots \\ a_{\pi_N} \end{pmatrix}.$$

A circular shift by one step is a special permutation

$$\boldsymbol{\Pi} = \begin{pmatrix} 0 & \cdots & 0 & 1 \\ 1 & \ddots & 0 & 0 \\ \vdots & \ddots & \ddots & \vdots \\ 0 & \cdots & 1 & 0 \end{pmatrix}.$$

For an l-step circular shift the permutation matrix is given by $\boldsymbol{\Pi}^l$.

B.3 Circulant matrices

An $N \times N$ circulant matrix is obtained by circularly shifting a column vector $\mathbf{a} = (a_0, \ldots a_{N-1})^{\mathrm{T}}$,

$$\mathbf{A} = \mathrm{circ}(a_0, \ldots, a_{N-1}) = \begin{pmatrix} a_0 & a_{N-1} & \cdots & & a_1 \\ a_1 & a_0 & & & a_2 \\ \vdots & \ddots & \ddots & \ddots & \vdots \\ a_{N-2} & & & a_0 & a_{N-1} \\ a_{N-1} & a_{N-2} & \cdots & a_1 & a_0 \end{pmatrix},$$

$$\mathbf{A} = a_0\mathbf{I} + a_1\boldsymbol{\Pi} + \cdots + a_{N-1}\boldsymbol{\Pi}^{N-1}.$$

Circulant matrices can be diagonalized as

$$\mathbf{A} = \mathbf{F}_N^{\dagger}\boldsymbol{\Lambda}\mathbf{F}_N = \mathbf{F}_N^{\dagger}\,\mathrm{diag}(\tilde{\mathbf{a}})\,\mathbf{F}_N,$$

where the eigenvectors are the columns of the DFT matrix $\mathbf{F}_N$ and the eigenvalues on the diagonal matrix $\boldsymbol{\Lambda}$ are given by the DFT of the vector $\mathbf{a}$,

$$\tilde{\mathbf{a}} = \mathbf{F}_N\mathbf{a}.$$

B.4 Linear and circular convolutions

Given vectors $\mathbf{a} = (a_0, \ldots a_{M-1})^{\mathrm{T}}$ and $\mathbf{b} = (b_0, \ldots b_{N-1})^{\mathrm{T}}$ of length M and N the linear convolution $\mathbf{c} = (c_0, \ldots c_{M+N-2})^{\mathrm{T}} = \mathbf{a} * \mathbf{b}$ is a vector of length

$M+N-1$ with entries

$$c_n = \sum_k a_k b_{k-n}.$$

Given vectors $\mathbf{a} = (a_0, \dots a_{M-1})^{\mathrm{T}}$ and $\mathbf{b} = (b_0, \dots b_{M-1})^{\mathrm{T}}$, both of length M, the circular convolution $\mathbf{c} = (c_0, \dots c_{M-1})^{\mathrm{T}} = \mathbf{a} \circledast \mathbf{b}$ is a vector of length M with entries

$$c_n = \sum_{k=0}^{M-1} a_k b_{[k-n]_M}.$$

The circular convolution can be also written as

$$\mathbf{c} = \mathrm{circ}(a_0, \dots, a_{M-1}) \cdot \mathbf{b}.$$

The circular convolution can be computed efficiently in the frequency domain using the standard properties of the DFT

$$\mathbf{c} = \mathbf{F}_M^{\dagger} \boldsymbol{\Lambda} \mathbf{F}_M \mathbf{b} \longrightarrow \mathbf{F}_M \mathbf{c} = \boldsymbol{\Lambda} \mathbf{F}_M \mathbf{b} \longrightarrow \tilde{\mathbf{c}} = \boldsymbol{\Lambda} \tilde{\mathbf{b}} \longrightarrow \tilde{\mathbf{c}} = \tilde{\mathbf{a}} \circ \tilde{\mathbf{b}}.$$

B.5 2D transforms, doubly circulant block matrices, and 2D circular convolution

Image processing typically deals with two-dimensional (2D) signals to perform various types of linear filtering based on 2D convolution operations and various types of 2D transforms. As discussed in this book, the delay-Doppler channel is a 2D representation of the linear time-variant high mobility channel, hence it is convenient to process 2D signals, which can be described by an $M \times N$ matrix when discrete-time samples are considered. Given a 2D signal matrix $\mathbf{X}$, its 2D-DFT is defined as

$$\hat{\mathbf{X}} = \mathbf{F}_M \mathbf{X} \mathbf{F}_N,$$

i.e., an M-point DFT is applied to each column of $\mathbf{X}$ followed by an N-point DFT of each resulting row. The inverse transform 2D-IDFT is given by

$$\mathbf{X} = \mathbf{F}_M^{\dagger} \hat{\mathbf{X}} \mathbf{F}_N^{\dagger},$$

i.e., an M-point IDFT is applied to each column of $\mathbf{X}$ followed by an N-point IDFT of each resulting row. The transform pair can be written in the vectorized form[1] as

$$\mathrm{vec}(\hat{\mathbf{X}}) = (\mathbf{F}_N \otimes \mathbf{F}_M) \cdot \mathrm{vec}(\mathbf{X}) \quad \text{and} \quad \mathrm{vec}(\mathbf{X}) = (\mathbf{F}_N^{\dagger} \otimes \mathbf{F}_M^{\dagger}) \cdot \mathrm{vec}(\hat{\mathbf{X}}).$$

[1]Using the property $\mathrm{vec}(\mathbf{ABC}) = (\mathbf{C} \otimes \mathbf{A}) \cdot \mathrm{vec}(\mathbf{B})$ when $\mathbf{C}$ is a symmetric matrix.

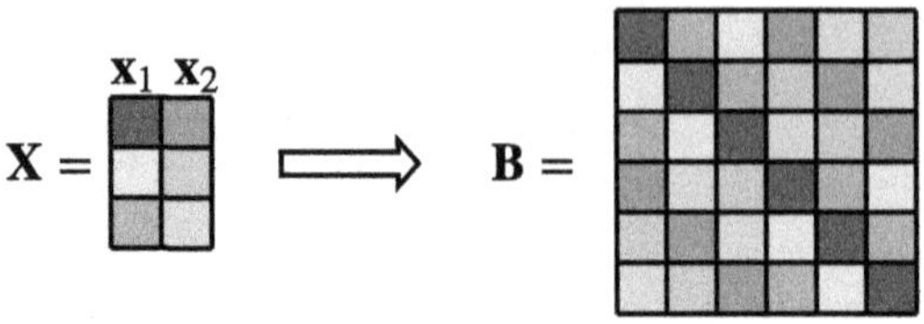

FIGURE B.1 Example of a doubly circulant block matrix $\mathbf{B}$ formed by $\mathbf{X}$ with $M = 3$ and $N = 2$.

We define the 2D circular convolution $\mathbf{Z} = \mathbf{X} \circledast \mathbf{Y}$ of $\mathbf{X} = [\mathbf{x}_1, \ldots, \mathbf{x}_N]$ with $\mathbf{Y}$, where

$$\mathbf{Z}[m,n] = \sum_{k=0}^{M-1} \sum_{l=0}^{N-1} \mathbf{X}[[m-k]_M, [n-l]_N]\mathbf{Y}[k,l].$$

In vectorized form we can write it as

$$\text{vec}(\mathbf{Z}) = \mathbf{B} \cdot \text{vec}(\mathbf{Y}),$$

where

$$\mathbf{B} = \text{circ}\,[\text{circ}(\mathbf{x}_1), \ldots, \text{circ}(\mathbf{x}_N)]$$

is an $MN \times MN$ *doubly circulant block matrix* (see Fig. B.1) which can be diagonalized as

$$\boldsymbol{\Lambda} = (\mathbf{F}_N^{\dagger} \otimes \mathbf{F}_M^{\dagger}) \cdot \mathbf{B} \cdot (\mathbf{F}_N \otimes \mathbf{F}_M),$$

where $\boldsymbol{\Lambda} = \text{diag}[\text{vec}(\hat{\mathbf{X}})]$. Then we can write

$$\hat{\mathbf{Z}} = \hat{\mathbf{X}} \circ \hat{\mathbf{Y}} \quad \text{or} \quad \text{vec}(\hat{\mathbf{Z}}) = \text{vec}(\hat{\mathbf{X}}) \circ \text{vec}(\hat{\mathbf{Y}})$$

to illustrate the convolution theorem for the 2D-DFT.

The *symplectic finite Fourier transform* (SFFT) is closely related to the 2D-DFT and is defined as

$$\hat{\mathbf{X}} = \mathbf{F}_M^{\dagger} \mathbf{X} \mathbf{F}_N,$$

i.e., an M-point DFT is applied to each column of $\mathbf{X}$ followed by an N-point IDFT of each resulting row. The inverse transform ISFFT is given by

$$\mathbf{X} = \mathbf{F}_M \hat{\mathbf{X}} \mathbf{F}_N^{\dagger}.$$

It can be easily shown that the above convolution theorem for the 2D-DFT equally applies to the SFFT.

APPENDIX

C

Some MATLAB® code and examples

C.1 Transmitter

MATLAB code 1 OTFS frame parameters.

```
% number of Doppler bins (time slots)
N=16;

% number of delay bins (subcarriers)
M=64;

% normalized DFT matrix
Fn=dftmtx(N);
Fn=Fn/norm(Fn);

% subcarrier spacing
delta_f=15e3;

% block duration
T=1/delta_f;

% carrier frequency
fc=4e9;

% speed of light
c=299792458;

% OTFS grid delay and Doppler resolution
delay_resolution = 1/(M*delta_f);
Doppler_resolution = 1/(N*T);
```

MATLAB code 2 Generate OTFS frame.

```
%modulation size
mod_size=4;

% number of information symbols in one frame
N_syms_per_frame=N*M;

% number of information bits in one frame
N_bits_per_frame=N*M*log2(mod_size);

% generate random bits
tx_info_bits=randi([0,1],N_bits_per_frame,1);

% QAM modulation
tx_info_symbols=qammod(tx_info_bits,mod_size,'gray','InputType','bit');

% Generate the MxN OTFS delay-Doppler frame
X=reshape(tx_info_symbols,M,N);

% Vectorized OTFS frame information symbols
x=reshape(X.',N*M,1);
```

MATLAB code 3 OTFS modulation.

```
Im=eye(M);

% row-column permutation matrix (Eq. (4.33))
P=zeros(N*M,N*M);
for j=1:N
 for i=1:M
  E=zeros(M,N);
  E(i,j)=1;
  P((j-1)*M+1:j*M,(i-1)*N+1:i*N)=E;
 end
end

% Method 1 (Eqs. (4.19) and (4.20))
X_tilda=X*Fn';
s=reshape(X_tilda,1,N*M);

% Method 2 (Eq. (4.35))
s=P*kron(Im,Fn')*x;

% Method 3 (Eq. (4.35))
s=kron(Fn',Im)*P*x;
```

C.2 Channel

MATLAB code 4 Channel mobility parameters.

```
% maximum user equipment speed
max_UE_speed_kmh=100;
max_UE_speed = max_UE_speed_kmh*(1000/3600);

% maximum Doppler spread (one-sided)
nu_max = (max_UE_speed*fc)/(c);

% maximum normalized Doppler spread (one-sided)
k_max = nu_max/Doppler_resolution;
```

MATLAB code 5 3GPP standard channels (from Section 2.5.1).

```
% EPA model
delays_EPA = [0, 30, 70, 90, 110, 190, 410]*10^-9;
pdp_EPA = [0.0, -1.0, -2.0, -3.0, -8.0, -17.2, -20.8];

% EVA model
delays_EVA = [0, 30, 150, 310, 370, 710, 1090, 1730, 2510]*10^-9;
pdp_EVA = [0.0, -1.5, -1.4, -3.6, -0.6, -9.1, -7.0, -12.0, -16.9];

% ETU model
delays_ETU = [0, 50, 120, 200, 230, 500, 1600, 2300, 5000]*10^-9;
pdp_ETU = [-1.0, -1.0, -1.0, 0.0, 0.0, 0.0, -3.0, -5.0, -7.0];
```

MATLAB code 6 Generate standard channel parameters.

```
% choose channel model, for example: EVA
delays=delays_EVA;
pdp=pdp_EVA;
% dB to linear scale
pdp_linear = 10.^(pdp/10);
% normalization
pdp_linear = pdp_linear/sum(pdp_linear);
% number of propagation paths (taps)
taps=length(pdp);

% generate channel coefficients (Rayleigh fading)
g_i = sqrt(pdp_linear).*(sqrt(1/2) * (randn(1,taps)+1i*randn(1,taps)));

% generate delay taps (assuming integer delay taps)
l_i=round(delays./delay_resolution);
```

```
% Generate Doppler taps (assuming Jakes spectrum)
k_i = (k_max*cos(2*pi*rand(1,taps)));
```

MATLAB code 7 Generate synthetic channel parameters.

```
% number of propagation paths
taps=6;

% maximum normalized delay and Doppler spread
l_max=4;
k_max=4;

% generate channel coefficients (Rayleigh fading) with uniform pdp
g_i = sqrt(1/taps).*(sqrt(1/2) * (randn(1,taps)+1i*randn(1,taps)));

% generate delay taps uniformly from [0,l_max]
l_i = [randi([0,l_max],1,taps)];
l_i= l_i-min(l_i);

% generate Doppler taps (assuming uniform spectrum [-k_max,k_max])
k_i = k_max-2*k_max*rand(1,taps);
```

MATLAB code 8 Generate discrete delay-time channel coefficients and matrix.

```
z=exp(1i*2*pi/N/M);
delay_spread=max(l_i);

% Generate discrete-time baseband channel in TDL form (Eq. (2.22))
gs=zeros(delay_spread+1,N*M);
for q=0:N*M-1
 for i=1:taps
   gs(l_i(i)+1,q+1)=gs(l_i(i)+1,q+1)+g_i(i)*z^(k_i(i)*(q-l_i(i)));
  end
end

% Generate discrete-time baseband channel matrix (Eq. (4.38))
G=zeros(N*M,N*M);
for q=0:N*M-1
  for ell=0:delay_spread
    if(q>=ell)
      G(q+1,q-ell+1)=gs(ell+1,q+1);
    end
  end
end
```

MATLAB code 9 Generate delay-time and delay-Doppler channel matrix.

```
% generate delay-time channel matrix (Eq. (4.55))
H_tilda=P*G*P.'

% generate delay-Doppler channel matrix (Eq. (6.1))
H=kron(Im,Fn)*(P'*G*P)*kron(Im,Fn');
```

MATLAB code 10 Generate r by passing the Tx signal through the channel.

```
% Method 1: Using the TDL model (Eq. (4.36))
r=zeros(N*M,1);
for q=0:N*M-1
 for ell=0:(delay_spread-1)
  if(q>=ell)
   r(q+1)=r(q+1)+gs(ell+1,q+1)*s(q-ell+1);
  end
 end
end

% Method 2: Using the time-domain channel matrix (G) (Eq. (4.37))
r=G*s.';

% Method 3: Using the delay-time channel matrix (H_tilda) (Eq. (4.54))
x_tilda=reshape(X_tilda.',N*M,1);
y_tilda=H_tilda*x_tilda;
r=P*y_tilda;

% Method 4: Using the delay-Doppler channel matrix (H) (Eq. (4.59))
x=reshape(X.',N*M,1);
y=H*x;
r=P*kron(Im,Fn')*y;
```

C.3 Receiver

MATLAB code 11 Add AWGN.

```
% average QAM symbol energy
Es = mean(abs(qammod(0:mod_size-1,mod_size).^ 2));

% SNR=Es/noise power
SNR_dB = 25;
SNR=10.^(SNR_dB/10)

% noise power
sigma_w_2=Es/SNR
```

```
% generate Gaussian noise samples with variance=sigma_w_2
noise = sqrt(sigma_w_2/2)*(randn(N*M,1) + 1i*randn(N*M,1));

% add AWGN to the received signal
r=r+noise;
```

MATLAB code 12 OTFS demodulation.

```
% Method 1 (Eqs. (4.24) and (4.27))
Y_tilda=reshape(r,M,N);
Y=Y_tilda*Fn;

% Method 2 (Eq. (4.35))
y=kron(eye(M),Fn)*(P.')*r;
Y=reshape(y,N,M).';

% Method 3 (Eq. (4.35))
y=(P.')*kron(Fn,eye(M))*r;
Y=reshape(y,N,M).';
```

MATLAB code 13 OTFS delay-Doppler LMMSE detection.

```
% vectorize Y
y=reshape(Y.',N*M,1);

% estimated delay-Doppler matrix (Eq. (6.18))
x_hat=(H'*H+sigma_w_2)^(-1)*(H'*y);

% QAM demodulation
x_hat=qamdemod(x_hat,mod_size,'gray');
```

MATLAB code 14 OTFS time domain LMMSE detection.

```
% estimated time domain samples (Eq. (6.19))
s_hat=(G'*G+sigma_w_2)^(-1)*(G'*r);

% MxN estimated delay-Doppler symbols (using Method 1 in code 12)
X_hat_tilda=reshape(s_hat,M,N);
X_hat=X_hat_tilda*Fn;
x_hat=reshape(X_hat.',N*M,1);

% QAM demodulation
x_hat=qamdemod(x_hat,mod_size,'gray');
```

C.4 Generate G matrix and received signal for OTFS variants

MATLAB code 15 RZP-OTFS.

```
z=exp(1i*2*pi/N/M);
delay_spread=max(l_i);

% Generate discrete-time baseband channel in TDL form (Eq. (2.22))
gs=zeros(delay_spread+1,N*M);
for q=0:N*M-1
 for i=1:taps
  gs(l_i(i)+1,q+1)=gs(l_i(i)+1,q+1)+g_i(i)*z^(k_i(i)*(q-l_i(i)));
 end
end

% Generate discrete-time baseband channel matrix (Eq. (4.38))
G_rzp=zeros(N*M,N*M);
for q=0:N*M-1
 for ell=0:delay_spread
  if(q>=ell)
   G_rzp(q+1,q-ell+1)=gs(ell+1,q+1);
  end
 end
end

% generate received signal after discarding CP
r=G_rzp*s.';
```

MATLAB code 16 RCP-OTFS.

```
z=exp(1i*2*pi/N/M);
delay_spread=max(l_i);

% Generate discrete-time baseband channel in TDL form (Eq. (2.22))
gs=zeros(delay_spread+1,N*M);
for q=0:N*M-1
 for i=1:taps
  gs(l_i(i)+1,q+1)=gs(l_i(i)+1,q+1)+g_i(i)*z^(k_i(i)*(q-l_i(i)));
 end
end

% Generate discrete-time baseband channel matrix (Eq. (4.83))
G_rcp=zeros(N*M,N*M);
for q=0:N*M-1
 for ell=0:delay_spread
  G_rcp(q+1,mod(q-ell,N*M)+1)=gs(ell+1,q+1);
 end
end
```

```
% generate received signal after discarding CP
r=G_rcp*s.';
```

MATLAB code 17 CP-OTFS.

```
z=exp(1i*2*pi/N/M);
delay_spread=max(l_i);
l_cp=delay_spread;

% Generate discrete-time baseband channel in TDL form (Eq. (2.22))
gs=zeros(delay_spread+1,N*(M+l_cp));
for q=0:N*(M+l_cp)-1
 for i=1:taps
   gs(l_i(i)+1,q+1)=gs(l_i(i)+1,q+1)+g_i(i)*z^(k_i(i)*(q-l_i(i)));
 end
end

% Generate discrete-time baseband channel matrix (Eq. (4.93))
G_cp=zeros(N*M,N*M);
for n=0:N-1
 for m=0:M-1
   for ell=0:delay_spread
     G_cp(m+n*M+1,n*M+mod(m-ell,M)+1)=gs(ell+1,m+n*M+l_cp+1);
   end
 end
end

% generate received signal after discarding CP per block
r=G_cp*s.';
```

MATLAB code 18 ZP-OTFS.

```
z=exp(1i*2*pi/N/M);
delay_spread=max(l_i);
l_zp=delay_spread;

% Generate discrete-time baseband channel in TDL form (Eq. (2.22))
gs=zeros(delay_spread+1,N*(M+l_zp));
for q=0:N*(M+l_zp)-1
 for i=1:taps
   gs(l_i(i)+1,q+1)=gs(l_i(i)+1,q+1)+g_i(i)*z^(k_i(i)*(q-l_i(i)));
 end
end

% Generate discrete-time baseband channel matrix (Eq. (4.109))
G_zp=zeros(N*M,N*M);
for n=0:N-1
```

```
  for m=0:M-1
    for ell=0:delay_spread
     if(m>=ell)
       G_zp(m+n*M+1,m+n*M-ell+1)=gs(ell+1,m+n*M+l_zp+1);
     end
    end
  end
end

% generate received signal after discarding ZP per block
r=G_zp*s.';
```

Index

D

E

F

Printed in the United States
by Baker & Taylor Publisher Services